Dr Frank Kirwan
and Professor J.W. McGilvray

Irish Economic Statistics

Institute of Public Administration, Dublin

First published 1983

by the Institute of Public Administration
59 Lansdowne Road, Dublin, Ireland

© F. Kirwan & J.W. McGilvray, 1983.

Cover: Detail from *Two tax-gatherers* by
Marinus van Reymerswaele (?1509-1567).
Reproduced by courtesy of the Trustees,
The National Gallery, London.

Designed by Della Varilly

Kirwan, Frank
    Irish economic statistics. — 2nd ed.
    1. Ireland — Economic conditions —
    1949 — Statistics
I. Title   II. McGilvray, James
330.9417'0824   HC260.5.A1

ISBN 0 906980 14 3

Photoset in 10/12 Baskerville by Printset & Design Ltd
and printed by Mount Salus Press, Dublin

Contents

Preface

The first edition of *Irish Economic Statistics* was published in 1968 and clearly fulfilled a need, for it was reprinted in 1969. Since that time there have been a number of major changes and innovations in the collection and presentation of Irish economic statistics, many arising through Ireland's accession to the European Economic Community in 1973. There is therefore a need for an updated and revised guide to the sources of Irish economic statistics and their methods of presentation and analysis, which the present publication endeavours to satisfy.

In undertaking this task the opportunity has been taken to extend the coverage of the book. There are now additional chapters on Income, Wealth, Expenditure and Taxation, Prices and Wages, and Regional Statistics, and the chapter on Distribution has been extended to include Transport and Communications. Other chapters, while retaining the same titles, have been substantially re-written. There is rather less emphasis on the historical development of particular economic statistics, and rather more on the use of economic statistics in contemporary economic research.

Like its predecessor, this edition is designed as a reference source as well as a textbook. To satisfy the former need, we have attempted to make each chapter self-contained, for the benefit of readers who may only be interested in a particular area of economic statistics. For this reason, systematic readers of the book will find some repetition in the discussion of sources and methods, particularly with regard to the definition and construction of index numbers. As a textbook, it is perhaps most appropriate to include a separate chapter on index numbers which would be read before turning to the chapters on specific areas of ecomic statistics. This however is not so convenient if the book is utilised as a general reference source.

The solution adopted here — for which no particular merit is claimed — is to incorporate the explanation of *price* index numbers, as used in agriculture, industry, foreign trade and so on, in a single chapter (Chapter 8: Prices and Wages). In contrast, explanation of the construction and interpretation of *volume* index numbers is included in each chapter in which the subject-matter (agriculture, industry etc.) involves the use of volume index numbers; thus Chapter 3 includes explanation of the construction of volume indexes of agricultural output, and Chapter 4 explains the construction of volume indexes of industrial output. A major reason for doing this is that there is a greater variety of methods of compilation of volume index numbers used in Irish economic statistics, so that a standard presentation of index number formulae is less easy for volume than for price indexes. Inevitably however this entails an element of repetition in the discussion of concepts and methods of volume indexes in different chapters, particularly in Chapters 3, 4 and 5. Nevertheless it is felt that this compromise best suits the aim of the book as both a textbook and reference source.

Our thanks are due to a number of anonymous readers, including staff of the Central Statistics Office, for constructive comments and suggestions and for identifying factual errors. Remaining errors and other failings are entirely the authors' responsibility. We also wish to thank James O'Donnell for his encouragement, patience and practical help; Isobel Sheppard, who typed the text and the innumerable revisions to it; Sarah O'Hara, who guided the book through its successive stages of editing, printing and proof-reading; and finally to our wives, Moira and Alison, for many reasons.

FXK
JmcG

1 Population and Vital Statistics

Population stands at the very centre of economic activity. From it stems not only the supply of labour but also the demand for the goods produced by these labour services. The structure of labour supply and the composition of consumer demand therefore reflect the underlying structure of the population. This underlying structure in turn influences the provisions which must be made for housing, education and health services for, as the age and sex composition of the population changes, so will the required mix of social policies. This chapter describes available demographic and vital statistics in Ireland and provides the basic tools for their analysis. The focus is on benchmark data of population size provided by the Censuses of Population and to a lesser extent *Labour Force Surveys*, and on time series data on births, deaths and migration.

Sources of population statistics

The most important sources of information about population in Ireland are 1. the *Census of Population,* 2. the biennial *Labour Force Survey* and 3. the *Reports on Vital Statistics.*

The Census of Population

A census aims at complete enumeration of the population of the country. Using this criterion the first reasonably satisfactory Census of Population taken in Ireland was that of 1841. Unofficial attempts had been made prior to this point to estimate the population of Ireland; these attempts are described and critically assessed by Glass and Taylor (1976) and by Connell (1950) and reproduced in facsimile form in Lee (1973). The first official attempt to undertake a census of population in Ireland was carried out over the period 1813 to 1815, following the 1811 census taken in the rest of the United Kingdom. The Irish census was never completed and was abandoned in 1815, its failure being attributed in part to the employment of anglicised Protestant enumerators amongst a predominantly Gaelic Catholic population. The 1821 and 1831 United Kingdom censuses were taken in Ireland, but the results are not considered satisfactory, in part because owing to poor question design the responses reflect the subjective biases of the enumerators to an unacceptable degree.

The problem was overcome in 1841 by the introduction of the householder's schedule, which has survived in largely unchanged format to the present day. In contrast to earlier censuses, where the enumerators secured their responses by interviewing the population, the 1841 and subsequent censuses[1] required the completion of forms by the householders themselves. A census of population was taken thereafter decennially until 1911 (see Thompson (1911)). The transition from British to Irish rule and the attendant political instability forced the cancellation of the planned 1921 census. The first census taken in the newly independent Irish state was in 1926 and subsequent censuses were undertaken in 1936 and in 1946. Nineteen fifty one saw a return to the practice of undertaking the census at the beginning of the decade, but a mid-decade census was also taken in the 1950s and the 1960s and was planned, but cancelled, for 1976. The cancelled census was replaced by an extremely limited census taken in April of 1979. The most recent comprehensive census was taken in 1981.

The censuses of population taken under British rule were frequently more detailed in Ireland than in the remainder of the United Kingdom and were surprisingly comprehensive. For example, the 1841 census requested information on sex, age, relationship to the head of the household, marital status and date of marriage, occupation, ability to read or write, birthplace, school attendance and housing conditions. In the absence of civil vital statistics or of parish register data the census also attempted to collect information on births, deaths and marriages during the preceding ten years. The 1851 census was even more detailed. It included questions on the use of the Irish language, on infirmities, on illness and on the number of idiots and lunatics. Subsequent censuses saw the gradual evolution of the householder's schedule. Religious profession was added in 1861 and remained in subsequent censuses (even though never asked in the remainder of the United Kingdom). The unemployed were enumerated in the censuses of 1881 and 1891.

Following the introduction of the compulsory registration of births and deaths in 1864, questions on these topics were discontinued, as was the question on date of marriage after 1871. The census of 1911, along with those of England, Wales and Scotland, included for the first time questions on fertility, in this case on the total number of children born alive to married women. The scope and coverage of census material on nineteenth-century Ireland is described in detail by Glass and Taylor (1976).

The process of evolution of the householder's schedule has continued in censuses to the present day, the range of questions rarely being identical from one census to the next. In 1971, for example, along with the standard information on name, age, date of birth, occupation, employment, etc. the householder's schedule also included questions on the usual residence one year prior to the census date, on the method of transport and distance travelled to work or school, on the age at which full-time education ceased, on the number of years' attendance at school or college and on any scientific or technical qualifications held. This information was sought for each individual within the household and additional questions were included on the amenities available to the household. Questions were asked on the number of rooms occupied, on the type of water supply and the availability of bath, shower and sanitary facilities and on the number of motor vehicles used by the household. The 1979 stopgap census consisted of only six questions, five of which related to basic demographic data, while the sixth requested information on change of residence from outside the state during the year ended 31 March 1979.

The results of the census appear in a varying number of volumes in the following years. The first to appear is the Preliminary Report, usually within four to six months of the census date. This is then followed by individual bulletins which are later aggregated into subject volumes. In the case of the 1971 census twelve such volumes were published. The subjects covered were:

1 Population by area, sex and conjugal condition
2 Industries
3 Occupations
4 Occupations and industries by age and conjugal condition
5 Housing
6 Household composition
7 Fertility of marriage

  8 Scientific and technical qualifications
  9 Irish language
  10 Usual residence, migration and birthplace
  11 Religion
  12 Transport and journey to work

In addition a considerable amount of unpublished cross-tabulated data
is available on request from the Central Statistics Office.

Apart from the actual census volumes, a useful summary source of
information is the *Statistical Abstract of Ireland* which, though nominally
annual, appears at somewhat irregular intervals. This publication
contains tables and notes derived principally from the census reports,
but in many cases brought up-to-date by current information on births,
deaths and marriages. The tables are prefaced by a brief though useful
commentary. In addition the *Irish Statistical Bulletin* occasionally
presents specific analyses of census data.

In all census publications the population described is what is known
as the 'de facto' population, i.e. all persons present within the
boundaries of the state on the census night, together with all persons
who arrive within the state on the succeeding morning not having been
enumerated elsewhere. Persons aboard ships or boats in port are
therefore included with the population, while those persons normally
resident in Ireland but temporarily abroad are excluded. The census is
normally taken at a point in the year, the month of April, when such
movements are assumed to be insignificant or at a minimum, so that the
census results relate fairly closely to the usually resident population.

The Labour Force Survey
The biennial Labour Force Survey (LFS) was introduced in 1975
following Ireland's accession to the European Economic Community.
Though the LFS is primarily a source of labour statistics, it is also a
useful secondary source of demographic data. Moreover, its results are
available at more frequent intervals than those of the Census of
Population. There are three crucial differences between the two data
sources. Firstly, the LFS is a survey rather than a census, and does not
aim at the complete enumeration of the population achieved in a
census. Only a proportion of the population is enumerated, typically
one in twenty or twenty-five, and the results grossed up to national
levels (the methodology of the Labour Force Surveys is described in

detail in Chapter 2). Secondly, whereas the Census of Population is taken on a de facto basis, the Labour Force Surveys are targeted on the usually resident population, and thus exclude those temporarily in the state while including those temporarily absent from the state at the time of the survey. By contrast, census of population results relate only to those who were actually present in the state on the night of the census. Thirdly, census results are secured through the medium of a householder's schedule completed by the respondent, while the Labour Force Survey results are secured by means of interviews.

The range of information sought in the Labour Force Surveys is quite comprehensive and includes occupation, industry in which employed, hours worked, secondary activity and employment situation one year previously. For unemployed persons questions are included on the duration of unemployment, the reasons for unemployment, and on the techniques employed in seeking employment. In addition, each survey normally includes supplementary questions on related topics. The 1975 Labour Force Survey included subjective questions concerning a person's assessment of his or her working environment, and its characteristics in terms of noise, hygiene and safety. Questions were also included on details of travel to work in terms of time taken, distance, means of transport and weekly expenditure on public transport.

The 1977 Labour Force Survey dealt with pensions, while that of 1979 covered vocational training and the transition from education to working life. Unfortunately, the range of results published from the Labour Survey Surveys has been somewhat less comprehensive. Five years after the taking of the first survey in 1975, the only results published relate to the age, sex, marital status and employment structure of the population. In part this reflects the evolving methodology of the survey, and the necessity to revise the results from 1975 and 1977 in the light of the 1979 Census of Population. The evolving methodology makes comparisons between successive surveys somewhat difficult, especially at sub-national level. Such matters are dealt with in detail in the report of the 1979 Labour Force Survey. At national level the composition of the population is finely disaggregated into five-year age bands by sex and marital status. However, by virtue of its being a survey, a similar degree of disaggregation is not possible for geographical areas below national level. Thus for the planning regions[2] population is disaggregated into five age bands, generally of fifteen years' length.

Reports on vital statistics
Neither of the sources discussed above, the Census of Population and
the Labour Force Survey, is designed to provide regular data on
fertility and mortality, two of the three components of population
change. Detailed information on these matters is provided in the form
of quarterly and annual vital statistics reports.

The *Quarterly Report on Births, Deaths and Marriages and on Certain
Infectious Diseases* contains data on births and deaths classified by county
of occurrence and of residence. In addition, deaths are cross-classified
by principal cause and by the county of residence of the deceased. At the
national level, the Quarterly Report presents time series data on a
quarterly basis over an eight-yearly period of the number of marriages,
births and deaths, and converts these to rates per thousand of the
estimated population. The deaths of infants under one year are
separately distinguished and expressed as a rate per thousand births.
The report also includes an estimate of the size of the national
population in the preceding April.

The *Annual Report on Vital Statistics* is published approximately three
years after the year of reference, and analyses in considerable detail the
data on marriages, births and deaths according to a wide variety of
classification systems. In the case of births, such events are classified by
the age and residence of the mother, by number of previous children
and by legitimacy status. The occurrence of multiple births is also
documented. Deaths are similarly classified by age, by sex, by
residence and by cause. The classification of marriages is by age, socio-
economic group, mode of celebration and residence of both parties.
Marriages where the intended residence of the couple is outside the
state are separately distinguished. On all three topics comparisons are
presented with other countries, typically with England and Wales,
Northern Ireland and Scotland. The statistical material is preceded by
a useful summary report.

Analysis of demographic data

Population structure
To analyse the changing structure of population over time, and to make
international comparisons of population structure, the data published
in the sources documented above must be transformed into readily
comparable form. There is no single summary measure which
adequately describes the structure and composition of a population, but

several techniques exist which isolate salient features. The most general method of describing the age and sex composition of a population is a graphical one and employs a device known as a population pyramid.

Figure 1.1: Population pyramid Ireland 1971

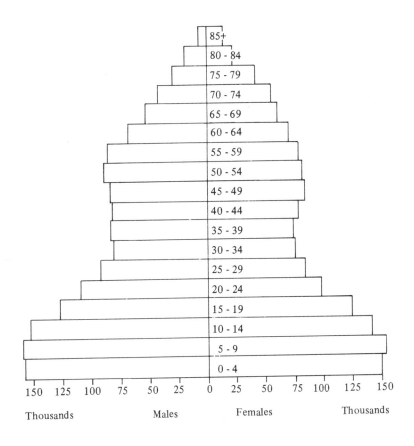

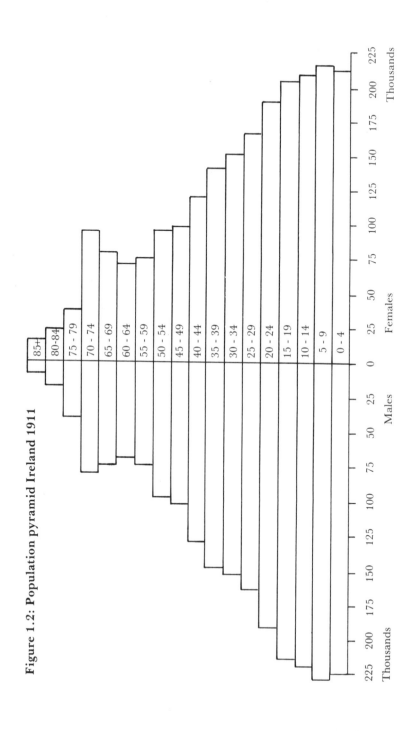

Figure 1.2: Population pyramid Ireland 1911

The population pyramid for Ireland, based on the 1971 census results, is presented in Figure 1.1. The technique of construction is as follows: age is scaled along the vertical axis, with the number of males and females being measured along the horizontal axis to the left and right respectively of the vertical axis. Each horizontal bar of the pyramid therefore represents a particular age group, the width of the bar being determined by the numbers actually present in that group. The resulting diagram is typically pyramidal in form, reflecting increasing mortality in later years. In cases where the population has been severely depleted, perhaps by means of war or emigration, this is readily apparent from the diagram. Population pyramids can, however, only be compared visually, there being no summary statistic to describe their structure.

However, a statistic which is readily comparable both across time and geographical boundaries is the dependency ratio. This measures the population in the dependent or non-productive age groups, typically taken as being those between birth and fifteen, and those aged sixty-five and over, as a percentage of the total population. A nation or region suffering sustained net emigration will typically have a high dependency ratio, reflecting the loss of the younger, more active members of the population. The dependency ratio is a relatively crude statistic, failing to take any account of the distribution of population within the active age groups.

A further useful comparative statistic is the density of population, which describes the number of people present in a particular area, typically that of a square mile or hectare. This measure must be interpreted with caution, as it takes no account of uninhabitable areas such as mountains, lakes and deserts.

Vital statistics

The analysis of population rests upon an examination of trends in deaths, births and migration and in the factors which affect them. Each of these three components of population change will now be analysed in turn.

Mortality The number of deaths occurring in a population over a specific time period is influenced by the age distribution of the population, its longevity and its sex distribution. The greater the proportion of the population in the upper age groups the greater the number of deaths to be expected, while the longer the average expectation of life of a given population, the fewer the number of deaths to be expected in a given time period[3]. Mortality is also affected by the proportion of males and females in the population, since in general females have a longer expectation of life than males.

Analysis of mortality starts with the number of deaths occurring in a specific time period; but this figure, being a function of the total population, is of limited usefulness and the normal practice is to relate deaths to population by means of the crude death rate. This rate is defined as

$$\frac{\text{Annual deaths}}{\text{Mean population}} \times 1,000$$

that is, deaths per thousand of the population[4]. Crude death rates for Ireland in recent years are shown in Table 1.1. Crude death rates are, however, likely to be misleading and their analytical uses are limited to de facto estimates of current changes in population or to forecasts of short term changes, where variations in the sex and age composition of the population are negligible.

Table 1.1: Crude death rate, Ireland 1864—1977

Period	1864-70	1871-80	1921-30	1961-70	1977
Crude death rate	16.23	18.06	14.45	11.66	10.28

Source: *Report on Vital Statistics 1977*, Table 5

A rising crude death rate over time does not necessarily imply that the health of the population or the expectation of life is declining. It may simply reflect the changing age composition of the population, resulting in a higher proportion of elderly people. Thus the crude death rate may well rise even though the death rate for any age group in the population is falling. International comparisons of crude death rates are equally hazardous. The distorting effects of age composition suggest that a more reliable indicator of the trend of mortality might be the calculation of a separate death rate for different age groups in the population, with separate rates for males and females to allow for changing sex composition. Such death rates are referred to as age-

specific death rates. For a particular age group the age-specific death
rate is defined as

$$\frac{\text{No. of deaths in age group x to x} + \text{a}}{\text{Mean population in age group x to x} + \text{a}} \times 1{,}000$$

where 'a' is the class interval of the specific age group. Such age-specific
death rates for Ireland, both for total population and for males and
females separately, are published in the *Annual Report on Vital Statistics*
and in the *Statistical Abstract*. Table 1.2 reproduces part of a recent table
from the former source showing the age-specific death rates for males
by ten-year age bands from 1871 to 1977. Examination of this table
reveals that, with the exception of those aged over seventy-five, age-
specific death rates have fallen during the past century, the falls being
most significant at lower ages.

Table 1.2	Age	1871-80	1921-30	1961-70	1977
Average annual	All ages	18.55	14.39	12.60	11.12
male death					
rates	Under 5	39.78	23.58	6.42	4.02
per 1,000	5—9	5.26	2.58	0.48	0.51
corresponding	10—14	3.05	2.58	0.48	0.36
population	15—19	5.08	3.21	0.74	0.78
by age	20—24	7.32	4.92	1.10	1.28
	25—34	8.88	5.33	1.26	1.10
	35—44	10.76	7.08	2.64	2.22
	45—54	15.38	11.44	7.32	6.87
	55—64	29.21	24.50	20.31	19.82
	65—74	66.54	48.08	49.66	48.72
	75—84	144.84	107.81	113.52	116.73
	85 +	298.18	217.55	287.11	273.47

Source: *Report on Vital Statistics 1977,* Table 9

Age specific death rates by cause may be calculated from the annual
Report on Vital Statistics which classifies deaths amongst the residents of
each province, county, and county borough by age, sex and cause of
death. The latter classification is by the abbreviated International
Standard List of Cause of Death, thereby facilitating international
comparison. The prime cause of death to persons aged under thirty-five

of both sexes is traffic accidents. Detailed statistics on this particular
aspect of mortality are to be found in *Road Accident Facts* published
annually by An Foras Forbartha.

Table 1.2 also reveals that the age-specific death rate for the group
'under five years' is high relative to subsequent quinquennial death
rates. An important component of the age-specific death rate for this
particular group is the infant mortality rate. This is defined as:

$$\frac{\text{Deaths of infants aged under one year}}{\text{Annual births}} \times 1,000$$

The infant mortality rate is an important indicator of the health and
welfare conditions of different countries, and of the trend in these
conditions. In recent years there has been a sharp fall in infant mortality
rates in many countries; this fall is illustrated in the case of Ireland in
Table 1.3. Infant mortality fell from 66 per 1,000 in the period 1941-50
to 16 per 1,000 in 1977.

Table 1.3	Period	Ireland	Northern Ireland	England and Wales	Scotland
Infant mortality	1941-50	66	60	43	57
rates in Ireland	1951-60	37	32	25	30
and neighbour-	1961	31	27	21	26
ing countries	1962	29	27	22	27
1941-1977	1963	27	27	21	26
	1964	27	26	20	24
Deaths of infants	1965	25	25	19	25
under one year	1966	25	26	19	23
per 1,000	1967	24	23	18	21
live births	1968	21	24	18	21
	1969	21	24	18	21
	1970	19	23	18	20
	1971	18	23	18	20
	1972	18	21	18	19
	1973	18	21	18	19
	1974	18	21	16	19
	1975	18	20	16	17
	1976	16	18	14	15
	1977	16	17	14	16

Source: *Report on Vital Statistics 1977*, Table XXXIV

The table reveals that a similar fall has taken place in the other countries within the British Isles, with the Irish fall being proportionately the greatest. Two further components of the infant mortality rate are also of interest: the neo-natal mortality rate and the peri-natal mortality rate. The former measures the number of deaths of infants aged less than four weeks per 1,000 live births while the latter measures the number of still births and deaths of infants aged less than one week per 1,000 live births plus still births. A somewhat less marked though still declining trend is apparent in these rates for Ireland and for many other countries.

None of the measures described provides a satisfactory overall means of comparing the level of mortality either inter-regionally or internationally. It is difficult to compare vectors of age-specific mortality rates for two different areas; infant and neo-natal mortality rates refer to only one aspect of mortality, and comparisons of crude death rates may be seriously misleading. Two measures employed in Ireland do not suffer from this deficiency: the mean expectation of life at birth (or its inverse, the Life Table death rate) and the standardised death rate. Each of these will now be described in turn.

A *Life Table* traces the mortality experience of a given cohort, typically of 100,000 persons, from birth until death. The method of construction of a Life Table is described in detail in an appendix to this chapter.

The basic procedure is as follows: the process starts in year 0 with a given cohort of births, usually 100,000, this cohort being subjected to given age-specific mortality rates as the survivors pass through each age group. Eventually the cohort dies off. The use of the age-specific mortality rates enables calculation of how many of the cohort will reach each age, or alternatively of how many will die before reaching that age. It is then possible to calculate the total number of years lived by the cohort which, when divided by the number in the cohort (i.e. 100,000), gives the mean expectation of life at birth. This figure, which may also be regarded as the average age at death of an individual in the cohort, provides an alternative measure of mortality which eliminates the effects of age distribution in the population. Similarly, the effects of the sex composition of the population may be eliminated by constructing separate life tables for males and females. In Ireland the differential mortality experience of urban and rural areas is accommodated by constructing separate Life Tables for males and females in each area. The Life Table therefore yields a statistic, the mean expectation of life at birth (or at any other age), which may be compared internationally without the distorting effects of differential sex and age composition.

Though the mean expectation of life at birth is frequently used as a means of comparing mortality experience internationally, this measure may also be transformed to yield an internationally comparable death rate. The death rate thus derived, called the life table death rate, is simply the reciprocal of the mean expectation of life, i.e. if X is the mean expectation of life at birth, then the life table death rate is 1/X multiplied by 1,000.

Life tables for Ireland, for both males and females, and for urban and rural districts, are generally published following each Census of Population. The publication outlet varies, the most recent being the Report on Vital Statistics 1974.[5]

It is important to emphasis that the Life Table and the measures derived from it are hypothetical, in that they assume certain specific mortality rates. In most cases the specific mortality rates are those of a particular year or recent number of years, though it is always possible to use purely hypothetical rates which attempt to account for possible changes in mortality conditions in the future. The most recently published set of Life Tables for Ireland uses mortality data for the years 1970-72. The results of the Census of Population for 1971 were used to standardise this data, hence the year to which the table refers is the census year. There are obvious drawbacks in using historical mortality data to predict future patterns of life expectation and mortality. For example, using the 1971 Life Table a child born in that year will be subjected in 1991 to the specific mortality rate for the age group twenty which prevailed in 1971. There may be good grounds for supposing that mortality conditions will be different in the future, and this is a factor to remember in using such tables. Nonetheless they are used for a wide variety of purposes, for example, by insurance companies and by economic and social programmers.

The second useful measure of comparative mortality is the *standardised death rate*. This rate is calculated by applying the age-specific death rates of a particular area to the total population and thereby obtaining a hypothetical figure for the total number of deaths which would occur if the national specific death rates were the same as those of a particular area. This hypothetical figure of deaths is then related to the total standard population used and the resulting death rate is called the standardised death rate for the particular area of interest. Standardised death rates for each county or urban district and for the four provinces are published in the *Annual Report on Vital Statistics*. As the tables in the report illustrate, it is quite possible for the crude death rate for a particular county to be well below the national crude rate but for the standardised death rate to be higher. It is the latter which best illustrates the comparative mortality conditions of an area.

Table 1.4 reproduces the crude and standardised death rates for the province of Leinster in 1977. A notable feature of this table is the extremely low crude death rate in Dublin county, reflecting the lower average age of population in that area. However, when correction is made for this atypical population structure and a standardised death rate calculated, the mortality experience of Dublin county no longer appears the most favourable: the standardised death rate is almost twice its crude level.

Table 1.4

Crude and		*Death Rates*	
Standardised	*Area*	*Crude*	*Standardised*
death rates	Carlow	10.78	11.35
in 1977	Dublin County Borough	9.65	10.80
	Dun Laoghaire Borough	12.00	9.81
	Dublin County (Remainder)	4.43	8.42
	Kildare	8.52	12.14
	Kilkenny	10.93	10.65
	Laois	10.70	10.45
	Longford	14.36	12.28
	Louth	8.89	10.13
	Meath	8.38	9.84
	Offaly	9.40	9.48
	Westmeath	11.14	11.08
	Wexford	11.21	10.60
	Wicklow	9.74	10.63
	Leinster	8.81	10.05
	National total	10.28	10.28

Source: *Report on Vital Statistics 1977*, Table XVIII

The method described above is the most direct way to calculate standardised death rates. Other methods are available. Reversing the procedure used above, for example, the age-specific death rates for the total or standard population could be applied to the population of a given area, giving rise to a hypothetical number of deaths which, when standardised by the area's population, yield a death rate known as the index death rate.

The index death rate will almost certainly differ from the standard and crude death rates for the selected area. For example, if the selected area contains a large number of older people then it is likely that the index death rate will be higher than the standard death rate. A comparison of the two rates indicates the extent to which the age distribution of the selected area differs from the age distribution of the whole population. The ratio of the standard death rate to the index death rate yields a quantity known as the comparative mortality ratio which may then be used to adjust the actual crude death rate for the particular area to eliminate the effects of age distribution, and to provide comparative measures of mortality for different areas. Once such comparative mortality ratios have been calculated for certain areas for a particular census year, they may then be used to adjust directly the crude death rates for subsequent years, on the assumption that there has been no marked change in the age distribution of the area. This obviates the necessity, implied by the direct method, of securing the actual age-specific death rates annually for each area, which may not be readily available.

Births and fertility The analysis of births, or fertility, is characterised by measures analagous to those used in the analysis of mortality.[6] There are, however, important conceptual differences between the measures as, unlike mortality, fertility describes an event which can happen more than once. Patterns of fertility, unlike those of mortality, will therefore reflect personal preferences and attitudes. Consequently, annual observations on fertility are likely to fluctuate much more widely than those on mortality, reflecting the ability to defer fertility temporarily —
in the circumstances of an economic downturn, for example.

The simplest measure employed in the analysis of fertility is the *crude birth rate,* that is, the annual number of births divided by the mean population multiplied by 1,000. The crude birth rate suffers all the defects of the crude death rate, being dependent upon the sex and age distribution of the population. It is therefore affected by the proportion of females of child-bearing age in the population and does not necessarily reflect the relative fertility of the population in question.

A better indication of fertility can be derived by relating the number of births to the number of females of child-bearing age in the population. The measure thus defined is the general fertility rate, calculated as annual births divided by the population of women of child-bearing age, multiplied by 1,000. This measure partially eliminates the effects of the sex and age distribution of the population, but does not take account of the distribution by age of females within the child-bearing span. To take account of this problem, age-specific fertility rates are calculated for each age or age-group of women. The specific fertility rate for a particular group is the annual births to females in that group divided by the number of females within the group multiplied by 1,000.

Such age-specific fertility rates can then be applied to a cohort of 1,000 females as they pass through the child-bearing ages of fifteen to forty-nine. The total number of children born to this cohort is summed, and then divided by the size of the cohort to yield the total fertility rate. This statistic measures the average number of children born to a female subject throughout her life to fertility conditions represented by the given vector of age-specific fertility rates. The resulting measure of fertility is hypothetical in that it assumes the continuance of certain fertility conditions generally based on average specific fertility rates for two or three recent years. Furthermore, the total fertility rate is based on age-specific fertility rates which will naturally be affected by the proportion of the female population who marry, and by other factors such as the age of marriage and the rate at which families are formed. The total fertility rate is subject to social and psychological influences which are liable to considerable variation so it tends to be unstable.

Specific fertility rates for Ireland are published in the *Annual Report on Vital Statistics*. The rates are calculated for five-year age bands within the child-bearing years; separate rates are calculated for all women and for married women only. Table 1.5 presents such age-specific fertility rates for 1960 and 1977, reflecting the changing pattern of fertility over this period. The Report on Vital Statistics does not include a measure of the total fertility rate, but this can be readily calculated from the published statistics. For example, the total fertility rate corresponding to the vector of age-specific fertility rates for 1977 in Table 1.5 is 3.25.

Reproduction A particularly important use of the mortality and fertility data described above is in the analysis of the reproduction of

Table 1.5: Age-specific fertility rates, 1960 and 1977

Age of mother at maternity	Rate per 1,000 women		Rate per 1,000 married women*	
	1960	1977	1960	1977
15—19	8.9	21.1	734.7	566.4
20—24	101.5	129.8	524.2	364.7
25—29	205.0	204.9	437.7	283.8
30—34	217.9	166.1	329.4	196.7
35—39	157.8	93.7	221.9	108.8
40—44	54.2	33.0	74.7	39.2
45—49	4.1	3.0	6.0	3.8
Total fertility rate	3.75	3.25	—	—

*legitimate births only.

Source: *Reports on Vital Statistics 1960 and 1977.*

population. The simplest measure of reproduction is given by the annual natural increase in population, that is, the difference between births and deaths during the year in question. This figure is generally expressed as a rate, the crude rate of natural increase, defined as annual births minus annual deaths divided by the mean population multiplied by 1,000. This formula represents the simple difference between the annual crude birth and death rates. In view of the shortcomings of these components of the crude rate of natural increase, this rate is likely to be a reliable guide to the rate of reproduction of the population only in the short run, when changes in the sex and age distribution are likely to be relatively insignificant.

The measure described above does not answer the crucial question of whether the population is reproducing or replacing itself over time. The relevant quantity in this case is not the total number of children born to a given cohort of women but rather the number of female children born to that cohort. The important factor is whether the female population is replacing itself or not, for in the long run this determines the trend of total population. The gross reproduction rate is the average number of female children born to a female throughout her child-bearing years. It therefore represents the female part of the total fertility rate and may be calculated from the latter. The gross reproduction rate is derived by scaling the total fertility rate by the proportion of births which are female.[7]

The gross reproduction rate thus defined does suffer some drawbacks, reflecting the fact that the total fertility rate on which it is based takes no account of female mortality during the child-bearing years. The net reproduction rate takes account of this problem. Calculated in a manner analogous to the gross reproduction rate, it subjects a given cohort of 1,000 women not only to prevailing age-specific fertility rates but also to mortality rates. The net reproduction rate therefore measures population replacement.[8] If the rate is equal to unity then the population is in the long run, and assuming no change in specific fertility and mortality rates, just replacing itself. A net reproduction rate in excess of (or below) unity indicates an eventual increase (or decline) in population. Even though in the short run population may be increasing, if the net reproduction rate is less than unity the population will eventually decline. The converse is also true, though both predictions are subject to the qualification that the factors which determine the specific fertility and mortality rates remain constant. This is a somewhat hazardous assumption, and consequently over-much reliance should not be placed on net reproduction rates as indicators of future population aggregates.

The net reproduction rate defined above refers to the replacement or turnover of a generation of females rather than to the annual change in the level of population. Thus, a value of, say, 1.2 for the net reproduction rate implies a ratio of 1,200 females to 1,000 from one generation to the next. Given the length of a 'generation' in years the net reproduction rate (R) may be expressed as an annual rate since $(1 + r)^n = R$ where 'n' is the length of the generation in years, 'R' is the net reproduction rate, and 'r' is the annual rate of increase which may be calculated given the values of 'n' and 'R'.

The derived annual rate is called the true rate of natural increase. Under the assumptions of constant fertility and mortality conditions it purports to measure the annual rate at which population will grow. Being a simple function of the net reproduction rate, the true rate of increase is both hypothetical and subject to the same limitations as that rate. Both rates relate to changes in the population of females in the child-bearing years, and do not take account of changes in the mortality conditions of females outside this band or of the male population, though such changes would clearly affect the rate of increase of the population.

The annual Reports on Vital Statistics do not contain any estimates of reproduction rates or population trends based on such rates. The census of 1946, 1961 and 1971 included inquiries on the fertility of marriage, but the published reports do not include calculations of rates of reproduction or replacement of population. However, the general report covering the 1946 and 1951 censuses contained a section on fertility of marriage which included estimates of gross reproduction rates for certain years between 1870 and 1950, and an estimate of the net reproduction rate based on the period 1950 to 1952.

Migration Population change over time results from three forces: mortality, fertility and migration. In most economies the first two factors tend to be more significant than the last. Not so in Ireland. Over the period 1841 to 1961 the loss of population from Ireland through external emigration exceeded the natural increase in each intercensal period,[9] with the exception of 1946-51.

In discussing migration it is necessary to distinguish between emigration, the flow of population from the country, and immigration, the flow of population into the country. The balance of these two gross flows yields the net migration flow. The term migration is not merely confined to movements across national boundaries but refers also to movement within a country, internal migration. Such migration leads to the inter-regional redistribution of population, with important influences on the demands for housing and social infrastructure.

The total rate of increase of the population is equivalent to the rate of natural increase plus the rate of net migration. In most countries the former far exceeds the latter. This is not the case in Ireland where, as noted above, the rate of net migration was significantly negative over the period 1841 to 1961 (with the exception of the post-war years 1946 to 1951) and numerically in excess of the rate of population increase over the whole of this period. During most of the 1970s, however, net migration appears to have been positive. Migration is the most volatile of the three components of population change, yet in Ireland it is also the most difficult to document. Partly as a consequence of this, official postcensal estimates of net external migration have not been published on a regular basis since 1949, and in the interval between successive censuses of population the size of the net flows can only be estimated by comparing successive official population estimates and correcting for the known numbers of births and deaths. The annual estimates of net migration yielded by such an exercise over the period 1926 to 1975 are presented by Hughes (1977), and suggest that with the exception of the depression year of 1932, and of 1940, the year following the outbreak of World War II, net emigration took place in every year between 1926 and 1971. The decade of the '70s was, however, characterised by net immigration.

The technique employed by Hughes (1980) can replicate the estimates of net migration implicit in the annual official population estimates, but can shed no light on how these net migration estimates are in fact arrived at by the Central Statistics Office. The official methodology has never been explicitly elaborated but Keating (1977) states that the official estimates are based on net passenger movements into and out of the state. Such passenger movements by rail, road, sea and air are documented monthly in the Irish Statistical Bulletin and at greater length in the Annual Statistical Abstract. Gross flows in both directions are identified, with three destinations specified — Great Britain, Northern Ireland and Other Places. By means of regression analysis Hughes (1980) has replicated the implied official estimates of net migration over the period 1946 to 1971 using this series. The use of such a methodology would explain the significant downward bias of the postcensal population estimates revealed by the 1979 census, as the historical relationship between net migration and net passenger movement appears to have collapsed in the years following 1971.

The *Manpower Information Quarterly,* published by the Department of Labour, documents the numbers of inter-regional migrants availing of the Resettlement Assistance Scheme. The origin of each migrant is classified by planning region for internal migrants, and as Great Britain and Northern Ireland, or Rest of the World, for those entering the country. The migrants' destination is classified by planning region. The series is annual and distinguishes five major occupational groups. The numbers thus assisted are small, and the series does not include data on family members accompanying migrants. Eligibility for assistance is means tested, so trends in the numbers availing of it reflect changes in eligibility as well as fluctuations in the volume of migration.

Details of the number of aliens employed annually under work permits since 1977 are published in the Manpower Information Quarterly. The total number of permits outstanding is classified by the alien's country of origin and by main economic activity. The numbers of new permits issued each year and the numbers renewed are similarly

documented. It must be remembered that EEC nationals coming to
work in Ireland have not required work permits since January 1978.
Furthermore, like those documented under the Resettlement
Assistance Scheme, the series provides no information on flows of
accompanying dependants.

Some further evidence on the size of migrant flows can be gleaned
from the periodic censuses of population, both of Ireland and of the
countries of destination of Irish emigrants. Hughes and Walsh (1975)
analysed the characteristics of Irish emigrants resident in Britain in
1971 using special tabulations from the 1971 United Kingdom Census
of Population. A similar study using data from the British *Labour Force
Survey* of 1977 has been carried out by Kirwan (1982), while O'Grada
(1975) used the results of post-Famine British censuses to provide an
estimate of emigration from Ireland over the period 1841 to 1911.

The Irish census of 1971 was the first to ask a question on residence
one year prior to the census date, and the responses to this question
provided the material for a study of internal migration in Ireland by
Hughes and Walsh (1980). An earlier study of internal migration in
Ireland was conducted by Geary and Hughes (1970) based on the birthplace statistics of the 1946 and 1961 censuses.

In the Irish context net migration is a particularly important
component of population change. The validity of postcensal population
estimates therefore hinges to a considerable degree on the adequacy of
the estimates of net migration. The downward bias of the official
postcensal estimates over the period 1971 to 1979 appears to have been
due in large part (Hughes (1980)) to a failure to model this component
adequately. Analytical research has proceeded along two lines with a
view to remedying this deficiency. The first of these, borrowing from
the burgeoning international literature on migration, has attempted to
specify econometric models of migration. In such models migration is
hypothesised to be a rational reaction to differentials in real incomes or
in labour market conditions. Studies which have followed this approach
have been those of Walsh (1974a), Geary and McCarthy (1977) and
Keenan (1978). While all these studies produced equations that fitted
the sample period reasonably well, the results were of little use for
predictive purposes. Given that those models generally used the
implied net migration estimates which have proved unsatisfactory in
the past, their poor forecasting performance is not surprising.

The second approach has attempted to infer migration movements
from changes in the numbers on the electoral register, after adjusting

for changes in the voting age and making allowances for the numbers of school-going children. This methodology has been developed for Ireland by Whelan and Keogh (1980) whose postcensal estimates for the period 1971 to 1979 did not suffer from the downward bias characterising the official estimates. However, the application of this model to the period 1979-81 has produced less encouraging results.

It is crucially important for the purposes of future economic and social planning that more reliable data on migration should become available and that less secrecy should attend the official estimation of net migration movements. Census of population data reveal that migrants are primarily in the younger, more active, age groups, a characteristic highlighted by the studies of Hughes and Walsh (1975, 1980). The distribution and magnitude of migrant flows will consequently have a significant effect upon longer term trends in population.

Population estimation and forecasting Postcensal estimates of the population of Ireland in June of each year are published in the *Quarterly Report on Births, Deaths and Marriages and on Certain Infectious Diseases.* The methodology involves taking the previous year's population estimate, adding the number of births in the current twelve month period, subtracting the number of deaths and making an adjustment for the level of net migration. Unfortunately, the CSO has never set out the process by which the latter component is estimated.

While population estimation involves trying to produce a figure for the size of the current population, population forecasting attempts to predict the size and structure of the population at some future point. The methodology of such population forecasting is simple and well established[10]. The number of persons aged X in any year t ($P_t(x)$) will be determined by those aged (X-1) in the previous year (t-1) who survive the full year $|P_{t-1} (x-1) - D_{t-1}(x-1)|$, where D_{t-1} (x-1) represents deaths in the previous year of those aged (x-1), plus or minus any net migration ($NM_{t-1}(x)$) in the intervening year of those aged X. Algebraically

$$P_{t(x)} = P_{t-1}(x-1) - D_{t-1}(x-1) + NM_{t-1}(x) \qquad (1)$$

This method cannot be used to predict the number of births, i.e. those aged 0 in mid-year t. Such births are determined by the number of women in the child-bearing years, taken in Ireland as fifteen to forty-nine, and by their propensity to bear children as measured by age-specific fertility rates. Algebraically

$$B_t = \sum_{i = 15}^{49} P_{ti} \cdot f(i) \tag{2}$$

Where B_t is the number of births in year t
P_{ti} is the number of females aged i in year t
$f(i)$ is the age specific fertility rate for age group i

The process represented by equations (1) and (2) is then repeated sequentially until the desired target year is attained.

The forecast population will obviously reflect the particular assumptions that are made about mortality, which is typically assumed to decline slowly to some specified level, fertility, which in Ireland is generally held to be falling, and migration which, as noted above, is extremely volatile. The results of this type of forecasting exercise are to be found in Walsh (1975) and Keating (1977). Such projections are extremely important in formulating future plans for economic and social development; Walsh's estimates were used to assess the implications for jobs and living standards (NESC (1975a)), for dwelling needs (NESC (1976a)) and for educational requirements (NESC (1976)).

Appendix

Notation and construction of life tables
The general method of construction of life tables has been explained in the main text. This appendix sets out the explanation in more detail in conjunction with an abstract from the Irish Life Table 1970-1972 for males living in urban areas.

The first row of the table begins with a cohort of 100,000 male births. The size of the cohort is shown in the column headed l_x where the subscript x denotes the age of each member of the cohort. Hence $l_0 = 100,000$. The third column, headed d_x, denotes the number of deaths occurring in the cohort between the ages x and x + 1, thus the first figure of 2,304 in this column reflects the number of deaths in the cohort between age 0 and age 1. This hypothetical figure is based on the actual infant mortality between 1970 and 1972. Returning to the l_x column, deaths in the first year reduced the original cohort by

LIFE TABLE

Urban District			(Males)				1970-72 extract	

Age x	l_x	d_x	p_x	q_x	L_x	T_x	e_x°	Age x
0	100,000	2,304	0.97696	0.02304	97,998	6,736,693	67.37	0
1	97,696	136	0.99861	0.00139	97,629	6,638,695	67.95	1
2	97,561	74	0.99924	0.00076	97,524	6,541,066	67.05	2
3	97,487	64	0.99934	0.00066	97,455	6,443,542	66.10	3
4	97,422	66	0.99932	0.00068	97,389	6,346,088	65.14	4
5	97,356	60	0.99938	0.00062	97,326	6,248,699	64.18	5
6	97,296	60	0.99938	0.00062	97,266	6,151,373	63.22	6
7	97,236	60	0.99938	0.00062	97,206	6,054,107	62.26	7
8	97,176	54	0.99944	0.00056	97,149	5,956,901	61.30	8
9	97,122	45	0.99953	0.00047	97,099	5,859,752	60.33	9
10	97,076	37	0.99962	0.00038	97,058	5,762,653	59.36	10
15	96,878	60	0.99938	0.00062	96,848	5,277,734	54.48	15
20	96,506	98	0.99899	0.00101	96,458	4,794,203	49.68	20
21	96,409	103	0.99893	0.00107	96,357	4,697,745	48.73	21
22	96,305	105	0.99891	0.00109	96,253	4,601,388	47.78	22
23	96,200	105	0.99890	0.00110	96,147	4,505,136	46.83	23
24	96,095	102	0.99894	0.00106	96,044	4,408,988	45.88	24
25	95,993	98	0.99898	0.00102	95,944	4,312,945	44.93	25
26	95,895	97	0.99899	0.00101	95,846	4,217,001	43.98	26
27	95,798	102	0.99894	0.00106	95,747	4,121,155	43.02	27
28	95,696	106	0.99889	0.00111	95,643	4,025,408	42.06	28
29	95,590	112	0.99883	0.00117	95,534	3,929,765	41.11	29
30	95,478	119	0.99875	0.00125	95,418	3,834,231	40.16	30
40	94,068	303	0.99678	0.00322	93,917	2,885,707	30.68	40
50	89,045	801	0.99101	0.00899	88,645	1,966,353	22.08	50
51	88,245	879	0.99003	0.00997	87,805	1,877,708	21.28	51
52	87,365	973	0.98886	0.01114	86,878	1,789,903	20.49	52
53	86,392	1,073	0.98758	0.01242	85,855	1,703,024	19.71	53
54	85,319	1,184	0.98613	0.01387	84,727	1,617,169	18.95	54
55	84,135	1,299	0.98456	0.01544	83,485	1,532,442	18.21	55
56	82,336	1,415	0.98291	0.01709	82,128	1,448,957	17.49	56
57	81,421	1,587	0.98125	0.01875	80,657	1,366,828	16.79	57
58	79,894	1,635	0.97953	0.02047	79,076	1,286,171	16.10	58
59	78,259	1,735	0.97783	0.02217	77,391	1,207,095	15.42	59
60	76,524	1,840	0.97596	0.02404	75,604	1,129,704	14.76	60
70	51,822	3,135	0.93950	0.06050	50,254	476,510	9.20	70
80	21,352	2,654	0.87568	0.12432	20,024	114,405	5.36	80
90	2,996	767	0.74407	0.25598	2,613	8,687	2.90	90
100	44	20	0.53927	0.46073	34	68	1.55	100
101	24	11	0.51479	0.48521	18	34	1.45	101
102	12	6	0.48959	0.51041	9	16	1.34	102
103	6	3	0.46366	0.53634	4	7	1.21	103
104	3	2	0.43700	0.56300	2	3	1.03	104
105	1	1	0.40962	0.59038	1	1	0.70	105

Source: *Report on Vital Statistics 1974*, Table 8a

Note: Only a selection of years from 0-105 appear here for space reasons

2,304, leaving 97,696 survivors at age 1. Hence $l_1 = 97,696$. It is important to grasp that this figure relates to the number of the cohort who survive to the exact age of 1, i.e. who are still alive on the first birthday. The columns l_x and age x must not be interpreted as representing a frequency distribution showing the numbers of persons within each age group. The values l_x refer to the size of the cohort at specific points in time, that is at the exact age x.

The fourth column of the table, headed P_x, is an empirical measure of the probability that a male aged x will survive to age x + 1. Thus the first figure in this column, 0.97696, measures the empirical probability of a male birth surviving the first year of life. This probability measure is in fact the ratio of the number who survive to age x + 1 to the numbers surviving at age x (in this instance the ratio of 97,696 to 100,000 or, in general the ratio $l_{x + 1}$ to l_x). The fifth column, q_x, is simply the converse of the fourth; it measures the probability that any member of the cohort aged x will not survive to age x + 1. It is therefore the ratio of the number of deaths between ages x and x + 1 to the number of survivors at age x. The first figure in the column, 0.02304, is the ratio of 2,304 to 100,000; or, in general

$$q_x = d_x/l_x$$

Since any member of the cohort must either die or survive between ages x and x + 1, $p_x + q_x$ must equal unity, and consequently q_x must equal $1 - p_x$. The sixth column, L_x, is a measure of the total number of years lived by the cohort between age x and age x + 1. If, for example, there were no deaths at all in the first year then the total number of years lived by the cohort from birth to age one would be 100,000, and L_0 would be 100,000. In practice, a proportion of the cohort will live the full year and the remainder, i.e. the d_x, for only part of the year. The simplest assumption to make is that, on average, each of the d_x will live for half a year; for example, if 100,000 of the cohort started the year and 1,000 died during the year, the total number of years lived by the cohort between age x and age x + 1 will be

$$l_x - d_x/2 = L_x$$

In the example taken here:

$$100,000 - (1,000)/2 = 99,500$$

Alternatively, $L_x = $ half $(l_x + l_{x + 1})$. One important exception to this rule, however, occurs in computing the value L_0. It is known that a large proportion of infant deaths occur in the first months of life, so that in this case it is unrealistic to assume an average life of six months for those who die between 0 and age one, and a somewhat lower average figure, roughly two months, is used for this group. In the actual table presented, $L_0 = 97,998$, and this consists of 87,696 years lived by those of the cohort who survived the first year plus an estimated 302 years lived by those 2,304 infants who died during the year. The latter estimate is based upon a detailed analysis of infant mortality statistics, by considering the probability of deaths at age of under one month, 1-2 months, 2-3 months, 3-6 months, and 6-12 months.

Ignoring for the moment the last two columns, the second row of the table denotes the experience of the cohort, of which there are now 97,696 survivors, between the ages one and two. Using the given specific mortality rate, 136 of them will die during the year, and the values for p, q and L follow from this. The process is continued until the hypothetical cohort dies off. The figures in the table, and the measures derived from it, are entirely dependent upon the specific mortality rates used, for these determine the values of d_x and hence all the other values in the table.

Turning now to the remaining columns, that headed T_x measures the total number of years lived by the cohort from the age of x upwards. It is therefore derived by cumulating the figures in the preceding L_x column. For instance the first figure, T_0, is the total number of years lived by the cohort from birth, i.e. the sum of all the figures in the L_x column. The second figure, T_1, is the total number of years lived by the cohort from the age of one onwards, and is therefore the sum of all the figures in the L_x column except the first one. It is easily seen from the table

$$T_0 = \Sigma L_x = \quad 6,736,693$$
$$T_1 = \Sigma L_x - L_0 = 6,736,693 - 97,998$$
$$T_2 = \Sigma L_x - (L_1 + L_0)$$
$$\quad = \quad 6,736,693 - (97,629 + 97,998)$$

The object in calculating the T_x column is to derive the average expectation of life of the cohort at a given age, and this is given in the e_x^o column. The total number of years lived by the cohort from birth is given in the table as 6,736,693 (T_0), and there are 100,000 survivors of the cohort at age zero. So the expectation of life at birth e_0^o, or the average number of years lived by any member of the cohort, is

6,736,693 divided by 100,000 = 67.37 years. Similarly at age one there are 97,696 survivors of the cohort and they are expected to live for a total of 6,638,695 years: the average expectation of life at age one e_1^0 is therefore 6,638,695 divided by 97,696 = 67.95 years. In general therefore, $e_x^0 = T_x$ divided by 1_x. To take an example further down the scale $e_{50}^0 = 22.08$, which means that those survivors of the cohort at age fifty may expect to live on average another 22.08 years.

For comparative purposes the most important of e_x^0 is e_0^0 showing the average expectation of life at birth, which in the above table is shown as 67.37. Alternatively, as was explained in the main text, this may be expressed as a unique death rate (the true death rate) as

$$(1/e_0^0) * 1000$$

in this case as 14.84 per thousand. The average expectation of life at birth is actually lower than the expectation of life at age one, though after that point the expectation of life slowly declines. The reason for this is the relatively high infant mortality rate in the first year.

Notes to Chapter 1

[1]   The principal results of 19th-century Irish censuses of population are reproduced in the population volumes of the Irish University Press, *British Parliamentary Paper Series*.
[2]   After the passage of the Local Government (Planning and Development) Act, 1963 the government established nine regions for physical planning purposes.

Government Planning Regions

Region	Area
East	Dublin, Kildare, Meath and Wicklow
South-East	Carlow, Kilkenny, Tipperary SR, Wexford and Waterford
South-West	Cork and Kerry
Mid-West	Clare, Limerick and Tipperary NR
West	Galway and Mayo
North-West	Leitrim and Sligo
Donegal	Donegal
Midlands	Laois, Longford, Offaly, Roscommon and Westmeath
North-East	Cavan, Louth and Monaghan

[3]   The three factors are interrelated and also affected by changes in birth rates. For example, if birth rates are falling then the effects of high longevity will eventually be countered by the change in age distribution.
[4]   Since population varies throughout the year, the figure used is a 'mean' population, the simplest definition of which is the arithmetic mean of the beginning-year and end-year populations.
[5]   The construction of Irish Life Tables is described in detail in the *Report on Vital Statistics 1974* pp 159-169. Source publications for all Irish Life Tables are as follows:

Life Table	Publication
No 1	*Census of Population of Ireland, 1926,* Vol V (pt. 1).
No 2	*Census of Population of Ireland, 1936,* Vol V (pt. 1).
No 3	*Register of Population of Ireland, 1941.*
No 4	*Census of Population of Ireland, 1946,* Vol V (pt. 1).
No 5	*Census of Population of Ireland — General Report 1946 and 1951.*
No 6	*Irish Statistical Bulletin,* June, 1965.
No 7	*Census of Population of Ireland, 1971,* Vol II.
	Irish Statistical Bulletin, March, 1972.
No 8	*Report on Vital Statistics, 1974.*

[6]   It is important to distinguish fecundity, which refers to the capacity to bear children, from fertility, which refers to the actual reproduction of children.

[7]   Since male births usually exceed female, the gross reproduction rate will be slightly less than half the total fertility rate.

[8]   In practice, the net reproduction rate is of only limited use as a measure of replacement, because of the inherent instability of fertility rates and the historically high level of emigration from Ireland.

[9]   See Glass and Taylor (1976) and the emigration volumes of the Irish University Press *British Parliamentary Paper Series* for data on 19th-century flows. For an analysis of Irish migration in the first half of the 20th century, see the *Report of the Commission on Emigration and Other Population Problems (1953).*

[10]   See for example Shorter and Pasta (1974).

2 Manpower

The previous chapter was concerned with total population; this chapter is concerned with that part of the population which is directly concerned with the production of goods and services. In what follows, the term 'labour force' is synonymous with the census of population concept of the 'total gainfully occupied population' which includes those out of work, temporarily resident in institutions, hospitals, etc., military and naval personnel, employers, employees and self-employed persons.

The labour force can be separated into primary and secondary segments. The former consists of those people whose attachment to the labour market is complete throughout their working lifetimes, and the latter consists of those people whose attachment to the labour force is considerably less strong, and whose working lifetimes will be marked by periods of withdrawal from the labour force. The latter category appears to consist almost entirely of married women.

Information on manpower is available at regular intervals. The most important sources are the biennial *Labour Force Survey*, the *Census of Population* and the *Monthly Returns from the Live Register*. Useful manpower data can also be derived from the *Quarterly Industrial Inquiry*, the *Census of Industrial Production*, and the *Censuses of Distribution and Agriculture*. With the exception of the results of the Labour Force Survey, most of the data collected are published in the annual *Statistical Abstract* and/or in the *Irish Statistical Bulletin*. Since 1979 such data have also been published in the *Manpower Information Quarterly* of the National Manpower Service, Department of Labour.

This chapter is structured as follows: the principal sources of manpower statistics, namely:

1 The Census of Population
2 The Labour Force Survey
3 The Census of Industrial Production

4 The Quarterly Industrial Enquiry
5 The Census of Distribution
6 The Annual Agricultural Enumeration

are described in detail. The interpretation of the published data is then discussed, focussing on questions such as seasonal adjustment, the impact of changes in social insurance on recorded levels of unemployment, and the adequacy of registered unemployment rates as indicators of labour market excess demand.

Statistics of employment

The Census of Population
The population census is the most detailed and comprehensive analysis of the country's labour force. As noted in Chapter 1, information on occupation was first collected in the 1841 Census, so that the census volumes provide details of the Irish working population over a long period of years.[1]

Two broad types of classification may be made, occupational and industrial, though prior to 1926 this distinction is sometimes unclear. The former groups individuals according to the type of work they do, for example typist, mechanic, accountant, lorry driver, and the latter groups individuals according to the industry in which they work, for example agriculture, coal mining, textiles, transport. For certain specific types of activity these classifications are synonymous, for example coal miner or brewer, but the majority of occupations are common to more than one industrial sector. Though the two types of classification are clearly inter-dependent, the distinction is important from an analytical point of view: in order to formulate employment and regional policies and to specify and pursue socio-economic objectives, it is necessary to know not only the general manpower requirements of sectors of the economy, but also the occupational requirements. For example, lack of particular categories of skilled workers may act as a bottleneck to industrial development. Such analysis and planning can be facilitated by the construction of industry/occupation matrices, which show the occupational composition of employment in each industry. To date such comprehensive matrices have not been compiled in Ireland, though some of the results of the AnCO *Manpower Survey* (1976), the Labour Force Surveys and the 1971 Census of Population have been presented in this format. For a recent example of the application of this technique to Scottish data see Fraser and Moar (1981).

The census results classify the total gainfully-occupied population by sex, area, occupation, industry in which employed, age and conjugal condition. The agricultural community is also classified by rateable valuation of farm or holding. Since 1951 the census results have included a socio-economic classification of the population based on occupation and employment status, distinguishing employers, own account workers, employees, etc. Eleven socio-economic groupings and a residual category were distinguished in the 1971 results as follows:

1 Farmers and farmers' relatives
2 Other agricultural employment (including fishermen)
3 Higher professional
4 Lower professional
5 Employers and managers
6 Salaried employees
7 Intermediate non-manual workers
8 Other non-manual workers
9 Skilled manual workers
10 Semi-skilled manual workers
11 Unskilled manual workers
12 Unknown

The Census of Population is the most comprehensive source of statistics on manpower, since it aims to include every person present within the state on the night of the census. The census results provide bench-mark data, a reference point, which can be updated during intercensal periods using less exhaustive techniques.

The Census of Population suffers from one drawback, the delay involved in the publication of the results. Figures on the characteristics of the labour force and its sectoral distribution are likely to be somewhat out of date by the time of publication. For example, the industry and occupational results of the 1971 census began to appear in 1973. Obviously, the greater the time-lag between collection and publication the less the value of the data as a guide to the manpower situation at the time of publication. More up-to-date information at shorter intervals is required. Such supplementary sources will now be described.

The Labour Force Survey: methodology

As noted in Chapter 1, the Labour Force Survey was inaugurated in Ireland in 1975. Such surveys have been carried out biennially thereafter as part of a European Community-wide operation involving simultaneous participation by all member states. Important definitional differences between the Labour Force Survey and the Census of Population are set out in Chapter 1. The most important conceptual distinction is that whereas the census aims at complete coverage of the population the Labour Force Survey only attempts to secure data from a representative sample. The more representative the sample supplying information, the more reliable the survey results will be as guides to economy-wide trends. It is worthwhile, therefore, to set out the background to sample surveys in some detail.

The characteristics of the labour force in a particular state can be analysed by collecting data on each individual present, i.e. the census method, or alternatively by collecting data on a small representative sample, i.e. the survey method. The latter method has several attractions; not only is it considerably cheaper to execute a survey than to carry out a full census, but the results of a survey take considerably less time to process than the full results of a census. It is possible, given the small numbers of people from whom survey data are to be collected, to employ trained interviewers, thus increasing the scope of the questions which may be asked and possibly improving the accuracy of the responses.

The aggregate from which information is required is called the target population. In the case of the Labour Force Survey, the target population is the aggregate of persons who are gainfully occupied. Such persons can be identified by sampling the population at large and then eliciting information only from the gainfully-occupied members appearing in the sample. The sampled population in the case of the Irish Labour Force Survey was the aggregate of all private households and all institutions within the state. For the purpose of the survey a private household was defined as any one person or group of persons with common living arrangements, separately occupying all or part of a private house, flat, apartment or private habitation of any kind.

The chosen sample in 1979 comprised approximately 32,000 households, and 310 separate institutions.[2] The sample was chosen by means of a two-stage process, designed to secure a sample whose demographic characteristics most closely matched those of the state as a whole. The entire state was sub-divided into enumeration areas, each

containing on average about 300 households. Each enumeration area
was then assigned to one of the following nine categories:

1 Dublin County Borough
2 Dun Laoghaire Borough and suburbs
3 Cork County Borough (excluding suburbs)
4 Limerick County Borough (excluding suburbs)
5 Waterford County Borough (excluding suburbs)
6 Towns with a population of 10,000 or more (excluding environs)
7 Towns with a population of between 5,000 and 10,000 (excluding
   environs) within each planning region
8 Towns with a population of between 1,000 and 5,000 within each
   planning region.
9 Other areas (including environs and suburbs of boroughs other
   than Dublin and Dun Laoghaire) within each county

Such a classification procedure is known as stratification, and a sample
selected from such a population is known as a stratified sample. The
sample was extracted by selecting a proportion of the enumeration
areas in each stratum. This proportion is known as the sampling
fraction. It varied between strata, being highest in the urban areas
where there is a greater variability in many of the important labour
force and population characteristics and being lower in rural, more
homogeneous, areas. In 1979 the sampling fraction varied from one in
ten in rural areas to one in four in the urban areas. Once the sample of
enumeration areas is selected, the first stage is complete.

The next stage of sampling for private households consisted of
choosing a systematic sample of one private household in three within
the selected enumeration areas (one in six in the Boroughs in 1979).
This was accomplished using an up-to-date register of private
households compiled by the survey interviewers in the months
preceding the actual survey. The final sample of private households
numbered 32,000 — approximately one household in twenty-five in
the entire country.

The sample of enumeration areas chosen in the first stage was also
used to select a sample of institutional households. In the selected
enumeration areas all institutions were identified, and the number of
usual residents determined in each case. All institutions having more
than fifteen usual residents in the chosen enumeration areas were

selected for interview. Depending on the size of the institution a sample of between one in four and one in ten of the usual residents was actually interviewed. In the case of institutions containing fewer than fifteen residents, the sample was selected separately within each enumeration area on a similar basis to private households. In the 1977 and 1979 surveys a number of large institutions which were not in the selected enumeration areas were included in the sample, as their exclusion in 1975 was believed to have led to a biased estimate of the institutional population of the state. There were 310 institutions included in the 1979 Labour Force sample.

The next stage in the survey is to administer a questionnaire to each household or institution. In the case of the Labour Force Survey this task was carried out by a specially recruited force of interviewers and supervisors. The response to the 1979 survey was excellent; a response rate of over 95% was secured. The response rate from institutions was close to 100%.

The final stage in any survey is the grossing up of the survey results to the level of the target population. The survey results for each individual enumeration area were initially aggregated by scaling them by the inverse of the sample proportion. For example, if one in twenty households within an enumeration area were interviewed, the responses secured were scaled by a factor of twenty to gross them up to enumeration area level. This procedure was carried out for each enumeration area, and the results summed. The second stage in the grossing-up procedure was to scale these results to national level. The weights used for this purpose were derived from the 1979 Census of Population and consisted of the ratio of the de facto population of the state as a whole in 1979 to the aggregate population of the sampled enumeration areas in that year. A similar grossing-up procedure was applied to the institutional returns.

Manpower data from the Labour Force Survey
As noted in Chapter 1, the Labour Force Survey involves a number of significant differences in methodology and definition compared to the Census of Population. It is important to remember this when comparing data from the two sources. The Labour Force Survey in Ireland is still in the evolutionary phase and data from successive surveys should be compared with caution.[3]

For the state as a whole the Labour Force Survey provides estimates of the number of persons at work classified by industrial group, sex, age, and employment status. A similar classification is presented by broad occupational group. Finally, the estimated number of persons at work is cross-classified by broad industrial group, broad occupational group and sex.

At the level of the broad economic sector, broad industrial group, and broad occupational group, the labour force at the national level is disaggregated into persons at work and persons who are unemployed, having lost or given up their previous job. At the broad sectoral level these data are also disaggregated by sex.

The survey also provides information on manpower at the regional level. The estimated number of persons at work is classified by broad economic sector and by sex for each of the nine planning regions. (For a definition of planning regions, see footnote 2 of Chapter 1, and Chapter 10). Separate totals are presented for Dublin County. It is thus possible to compare the sectoral composition of the labour force across regions. The estimated number of persons at work in each planning region is also classified by broad occupational group and by sex.

In interpreting the results of the Labour Force Survey it must be borne in mind that they are based on a sample of approximately one in twenty households. The results are therefore characterised by a certain amount of sampling variability or sample error. The magnitude of this variability will be lowest for the largest and most broadly defined aggregates, such as the total labour force, and highest for the more narrowly defined aggregates, such as the number of persons in a particular occupational group in a particular region. For this reason the results of the Labour Force Surveys cannot be finely disaggregated with any confidence to the regional level.

The two foregoing sources, the Census of Population and the Labour Force Survey, are the only economy-wide sources of manpower information. With the exception of the Live Register Returns all other sources are sector-specific.

The Census of Industrial Production
The Census of Industrial Production (described in detail in Chapter 4) has been carried out annually since 1931 and is intended to cover all establishments employing more than three persons in the transportable goods industries, as well as certain service industries such as building and construction (public sector only), gas and electricity, water and laundries. The results of the census are published regularly in the Irish Statistical Bulletin in the form of individual industry reports and

summary tables. The results are also published with a greater time-lag in the Statistical Abstract.

The employment figures in the census reports relate to the number of persons engaged during a week in September of the census year. The detailed industry reports analyse persons engaged by sex, and distinguish between industrial workers, apprentices, supervisory staff, outside piece-workers and administrative, clerical and technical staff. The industrial classification system used up to 1973 was based on the United Nations International Standard Industrial Classification (ISIC) as it related to Irish conditions. The results for 1973 and subsequent years have been reclassified in accordance with the general classification of economic activities (NACE) used by the EEC. The figures for total employment in establishments covered by the Census of Industrial Production are not directly comparable with Census of Population estimates for employment in corresponding industries or groups of industries.

There are two reasons for this lack of correspondence. Firstly, the Census of Population enumerates all persons gainfully employed; the Census of Production, by contrast, excludes all establishments with fewer than three persons engaged. Secondly, there are differences in classification. Furthermore, the Census of Industrial Production collects information in respect of establishments and therefore permits a detailed classification of the persons engaged in each unit. The Census of Population, by contrast, classifies persons by means of information supplied by the individual himself. Such information is not always sufficient to allow for classification at the level of the production unit, so that the person may be classified to the activity of a more broadly defined enterprise of which the relevant production unit is only a part. In certain industries it is often difficult to identify all the establishments in the industry and to obtain census returns from them. Most industries are therefore affected to some degree by differences in classification and hence, both for individual industries and in aggregate, the Census of Production and Census of Population figures are not directly comparable. Changes in employment in Census of Production industries should be treated with caution as indicators of changes in the general level of employment or of trends of employment in particular industries.

The Quarterly Industrial Inquiry
The Quarterly Industrial Inquiry (described in Chapter 4 below), a

sample survey of firms in the transportable goods industries, was
introduced in 1942 to secure regular data on output trends. Since 1950
the Inquiry has also collected data on earnings, employment and hours
of work. Such data are collected for industrial workers, managerial,
technical, clerical and other employees (including working
proprietors). The sample covers approximately 1,900 establishments,
accounting for over 90% of employment in transportable goods
industries. In practice, all establishments employing more than twenty
persons are included. The inquiry is a voluntary one, unlike the Census
of Industrial Production; nonetheless, the response rate is extremely
high. The responses from firms in the sample are grossed up to yield an
estimate of employment in each industry. The grossing-up factors are
derived from the most recent Census of Industrial Production, and
comprise for each industry the ratio of total employment within that
industry to employment in the sampled firms within the industry,
allowance being made for employment in establishments commencing
operations since the census year. The results are published in the Irish
Statistical Bulletin, subject to revision in the light of later Census of
Production results. The inquiry has yielded monthly employment
estimates since 1973.[4]

The Census of Distribution
The Census of Distribution (described in detail in Chapter 9 below) is
the principal source of information on employment in the services
sector. Full censuses of distribution have been taken in respect of the
years 1933, 1951, 1956, 1966, 1971 and 1977. For the years 1952-1955,
1957-1960 and 1967-1970 sample censuses were carried out based in
each case on the preceding complete census. Beginning in 1982, it is
planned to introduce an annual sample survey of distribution. The
results of past enquiries are published both in the Irish Statistical
Bulletin and in separate reports on the Census of Distribution.

    The coverage of the 1971 census extended to all permanent business
premises engaged in wholesale and retail trade and in the provision of
certain services. The 1977 census was restricted solely to wholesale and
retail trade; it analysed those engaged in such trade by age, sex,
geographical location, type of business and status (proprietor, unpaid
member of family, full-time or part-time employee). Enterprises were
classified by number of persons employed, legal status of
establishment, number of establishments operated by the enterprise
and duration of present ownership.

The Annual Agricultural Enumeration
The Annual Agricultural Enumeration taken in June of each year is an important indicator of the trend of overall employment in the agricultural sector. This data source, described in greater detail in Chapter 3, provides annual estimates of the number of males engaged in farm work. The results are disaggregated by county, by employment status (member of family, permanent or temporary) and two broad age categories, those between fourteen and eighteen years, and those eighteen years and over. At national level, the results are published in the Irish Statistical Bulletin. The results of the most recent enumeration are generally accompanied by comparative figures for the previous twenty years.

Since 1975 the CSO has carried out an annual survey of permanent agriculture workers, documenting the age, sex and hours worked of the permanent agricultural work force for a selected reference week. The results are published in the Irish Statistical Bulletin. (For further details of this survey see Chapter 3 below).

Other sources of manpower statistics
Until the advent of the Labour Force Survey, data on employment outside the agricultural and industrial sectors were fragmentary. Censuses of distribution were taken at isolated intervals, remedying the data deficiency to some extent, at least as far as the services sector was concerned. Notwithstanding the advent of the Labour Force Survey, information on employment in the tertiary sector of the economy is much less detailed than that in manufacturing industry and agriculture.

Economic activity outside that covered by the Censuses of Production and Agriculture may be divided into public administration and defence, private construction, distribution and transport, and other services. Only the transport (see Chapter 9) and private construction sectors are characterised by regular manpower statistics. Data on the former relate almost exclusively to staff employed by CIE. The average number employed throughout the year is disaggregated by mode (railway/road) and by function (passenger/freight). Within each category staff are further classified as administrative or operative.

An annual *Census of Building and Construction* has been carried out since 1974.[5] It extends only to construction undertaken by private sector firms, that carried out in the public sector being included in the Census of Industrial Production. The results distinguish employment in such firms during a particular week in September by size of firm and type of employee (i.e. working proprietor, managerial, technical or clerical, apprentices and labour only sub-contractors). The CSO also carries out a monthly survey of employment in the construction industry, the results of which are published in the Irish Statistical Bulletin. The CSO warns that the results of both these exercises are unreliable, owing to the difficulty of identifying all firms engaged in this sector and the poor response rate from those identified.

The Department of the Public Service carries out an annual census of employment within both the civil service and the public sector, but the results remain unpublished.

The Industrial Development Authority has carried out a survey of employment in grant-aided manufacturing industry in January of each year since 1973. The firms covered have availed of IDA incentives in the past, and the results provide data on the regional distribution of manufacturing employment by sex. No disaggregation of the results by age, occupation or industry is available. The data are unpublished and are intended primarily for IDA internal use, though in recent years details of the net change in manufacturing employment revealed by the survey have been published in the Manpower Information Quarterly.

An Chomhairle Oiliúna (AnCO) has carried out surveys of employment in certain manufacturing industries and in building and construction in 1974, 1975, 1976 and 1979. The surveys cover those firms participating in the AnCO levy/grant scheme, in practice all firms employing more than ten to fifteen people. Detailed results are published[6] by planning region for eight broad industrial sectors, providing for each sector a breakdown of employment by broad occupational group. Within each sector, the occupational data for each region are further disaggregated by major industrial heading. Information is provided on the regional distribution of apprentices by occupation within certain industries. Seven separate trades are distinguished — furniture, printing, electrical, motor, engineering, building and dental. Within each trade the apprentice population is analysed by years of apprenticeship completed. This analysis is performed for the end of March and the end of September, and the results published in the Manpower Information Quarterly. The number of trainees at AnCO training centres is similarly documented on an annual basis in the same publication.

Since 1961 the Economic and Social Research Institute and the Confederation of Irish Industry have undertaken regular surveys of employers' attitudes and intentions in the Irish manufacturing sector.[7]

These surveys, initially quarterly, have been carried out monthly since Ireland's accession to the EEC as part of the community-wide Harmonised Business Survey. The survey results attempt to identify the extent to which firms are constrained by labour shortages,[8] separately distinguishing sex and skill level, and to gauge firms' employment intentions over the next three month period.

The number of redundancies notified by employers to the Department of Labour is published annually in the Irish Statistical Bulletin[9] and on a quarterly basis since 1975 in the Manpower Information Quarterly. The figures are classified by industry and sex. The interpretation of the figures is difficult, for several reasons. The series covers only 'qualified redundancies', excluding therefore employees with less than two years service, those under sixteen or over seventy and those working fewer than twenty-one hours a week. Furthermore, some redundancies notified may later be rejected as unqualified. It is financially attractive for employers to notify redundancies well in advance; consequently redundancies notified within a quarter may not take place within that quarter, or indeed at all, being withdrawn before they take effect.

The National Manpower Service has carried out an annual survey on the labour market experience of school leavers since 1975.[10] The previous year's leavers are classified by seven further education categories, five employment categories and others ranging from 'emigrated' to 'destination unknown'. The results of these surveys are published in the Manpower Information Quarterly. Up to 1979 the information was provided by teaching staff some months after the end of the school year, and may therefore to some extent reflect incomplete information held by respondents. Perhaps for this reason, not all schools participated. AnCO estimated that completed returns accounted for over 85% of school leavers. The survey methodology was changed in 1979, when a sample of 3,000 school leavers was interviewed, chosen from lists of leavers provided by school authorities.

Details of the number of aliens employed annually under work permits since 1977 are published in the Manpower Information Quarterly. The total number of permits outstanding is classified by the alien's country of origin and by main economic activity. The numbers of new permits issued each year and the numbers renewed are similarly documented. In analysing the data it must be remembered that EEC nationals coming to work in Ireland have not required work permits since January 1978.

The Manpower Information Quarterly documents the volume of payments under government-sponsored job creation and employment maintenance schemes, and under the Work Experience Programme. In each case the Quarterly documents the number of applications received from employers, the number of employers included therein, and the number in respect of whom government funds were disbursed, distinguishing between adults and school-leavers. The data are disaggregated into five broad industrial groups on both a quarterly and an annual basis. In the case of the Work Experience Programme, quarterly flows into and out of the scheme are also documented. The impact of such programmes on levels of employment and unemployment has been examined by Nolan (1978A).

The advent of the biennial Labour Force Survey has greatly increased the range of available manpower statistics, but there is no single comprehensive publication which draws together the various data sources listed above. The potential for such a publication exists in the *Trend of Employment and Unemployment,* but the content of this publication is disproportionately slanted towards unemployment statistics. With the wider availability of employment data in recent years this position can now be redressed. Indeed, to some extent, the Manpower Information Quarterly of the National Manpower Service, introduced in 1979, embraces most of the data sources described here.

Statistics of unemployment and vacancies

Statistics of unemployment are primarily obtained through the operation of the Social Welfare Acts and the Unemployment Assistance Acts administered by the Department of Health and Social Welfare. Since 1911, when the foundations of the present social insurance schemes were laid, there have been frequent legislative changes affecting the scope and administration of schemes of social welfare, and it is necessary to bear this in mind in analysing the related statistics over a period of years. Walsh (1974B) is particularly useful in this respect as is the *Report of the Interdepartmental Study Group on Unemployment Statistics* (1979). The most important changes are listed in an appendix to this chapter.

The raw data from which published unemployment statistics are compiled are the number of persons registered at each office of the Department of Social Welfare; this is called the Live Register. The total registered at all offices is called the Total Live Register. In theory those on the register are persons who are able and willing to work but are unable to find suitable employment. In practice, however, the number of persons who register is affected by the eligibility conditions defined in the appropriate Social Welfare and Employment Assistance Acts and by the real value of unemployment compensation payments (Walsh (1978)).

Statistics of the live register at each local employment office at the end of each month are published in the Irish Statistical Bulletin.[11] At national level those on the register are disaggregated by sex, by occupation, by type of claim, and by the industrial group in which they are normally engaged. At less frequent intervals the unemployed are disaggregated by number of dependents, by duration of continuous registration, and by age. The distribution of registrants by planning region at the end of each quarter since 1976 is documented in the Manpower Information Quarterly. These analyses are later reproduced in part in the Trend of Employment and Unemployment.[12]

Persons on the live register fall broadly into two categories, claimants for unemployment benefit and applicants for unemployment assistance. Entitlement to the former is dependent upon the social insurance contribution record of the claimant. The scope of the social insurance scheme has varied considerably over the years; up to 1974 certain categories of employees were excluded, notably those whose income was above a specified level and those employed by their own husbands and wives. As the income limit for eligibility within the scheme was adjusted only irregularly, the series documenting claimants for unemployment benefit tends to be somewhat lumpy. At present the scheme is comprehensive, covering all employees, though contribution rates and eligibility to benefits vary between classes of employee. Public service employees, who pay a reduced contribution, are not eligible for unemployment benefit.

Unemployment assistance, by contrast, is not insurance-related. In theory, all unemployed persons between eighteen and sixty-six, whether insured or not, are entitled to unemployment assistance during periods of unemployment, subject to their holding 'a qualification certificate'. Until 1978 this requirement was used to exclude some categories of women from assistance while preserving the entitlement of males with identical characteristics. Rates of assistance, which are lower than for unemployment benefit, vary according to area, number of dependents, and the means of the claimant.

The Unemployment Assistance Acts qualify for assistance certain persons who are not, strictly speaking, unemployed, but are rather under-employed, of whom the most important are small farmers and their assisting relatives. Smallholder applicants for unemployment assistance have not been included in the live register totals since 1966. In order to exclude this group from continuous assistance, two Employment Period Orders were issued in some years before 1971 by the Minister for Social Welfare. These excluded, for certain rural areas and certain periods of the year, payment of unemployment assistance to such categories of claimant. Consequently, historical fluctuations in the live register cannot be taken as a very accurate reflection of real changes in the employment situation over time, at least prior to 1971. For more recent years the adequacy of registered unemployment as a proxy for labour market excess demand is reduced by the fact that the propensity of the unemployed to register appears to be a function of the level and duration of benefits available. It is not surprising therefore that those for whom no benefits are provided, such as school-leavers awaiting their first job and married women who have exhausted their entitlement to unemployment benefit, do not register.

The propensity to register will depend not only on the availability of financial incentives, but also on the scale of these incentives. Walsh (1978) contends that the introduction of pay-related unemployment benefits in 1974 significantly increased the average duration of unemployment.[13] The extension in 1976 of the period over which both flat rate and pay-related unemployment benefits are payable would have had similar effects. For all the foregoing reasons statistics of unemployment derived from the live register must be interpreted with considerable caution.

Unemployment rates
The proportion of the labour force unemployed at any one time is an important guide to economic policy. Published unemployment rates provide the most common measure of the balance between demand and supply in labour markets. However, as noted by Taylor (1974), unemployment rates as generally defined are an imperfect proxy for the level of excess demand or supply in the labour market. This is especially the case in Ireland, where the published unemployment rate is obtained as the ratio of the number of currently insured persons on the live register to the total currently insured population. Since the proportion of the total labour force included within the insurance scheme has varied over time, problems arise in analysing a time series of rates based on such a denominator.

The published rate has two major deficiencies. Firstly, the numerator and denominator exclude by definition that part of the population, whether employed or unemployed, which is not insured. Secondly, the denominator of the ratio is the number of insurance cards exchanged in the year prior to the year to which the live register refers. Thus, in calculating unemployment rates for 1980, the denominator is the number of insurance cards exchanged during the calendar year 1979. The number of cards exchanged, insofar as it relates to any measure of the labour force, measures those insured persons who worked or who were on the live register in 1978. During a period when the labour force is changing rapidly, whether by natural growth of population or by immigration, the use of a denominator which can be up to two years out of date is likely to lead to a biased estimate of the unemployment rate. The calculated rate will exceed the true rate where the labour force is expanding, and understate the true rate where the labour force is contracting. The consequences of this mis-specification have been explored by Sandell (1974) and Geary and Jones (1975). The exchange of insurance cards ceased with the introduction of the Pay Related Social Insurance scheme in April 1979. Insurance records are now computerised, and the system can provide up-to-date estimates of the size of the currently insured population.

Vacancy statistics
It is common, though perilous, to use unemployment as a measure of the balance between demand and supply in the labour market. The unsatisfactory nature of this variable has led to consideration of the alternative use of vacancy statistics. Unfortunately, this option is not readily available in Ireland as the only published series[14] on vacancies refers to 'vacancies notified at and filled through National Manpower Service Offices and Local Employment Offices'. Before the introduction of the National Manpower Service these figures related mainly to vacancies for work financed from central government funds, and the number of such vacancies declined almost continuously from 1960 to 1971. Since then there has been an increase in the number of vacancies notified, but the coverage remains far from comprehensive.

The most detailed source of vacancy statistics is the *Manpower Information Quarterly,* which documents on an annual basis since 1973 and on a quarterly basis since 1978 the number of vacancies notified to and filled by NMS. The data are disaggregated by five major occupational groups. In addition this publication documents, for each quarter since the beginning of 1977: vacancies unfilled at outset of quarter, those notified during each quarter, those filled by the NMS during the quarter, those filled by other means or cancelled, and those unfilled at the end of the quarter. Since the beginning of 1978 these data have been disaggregated quarterly by the following five occupational groups: managerial, professional and executive; clerical and secretarial; service occupations; farming, fishing and horticulture; industrial occupations.

The same publication documents the number of job seekers registered each quarter since 1977 with the NMS, again subclassified by five occupational groups. A detailed occupational breakdown of registered job seekers by sex at the end of each quarter has been published since March 1978. Flows onto and off this register are recorded quarterly since the beginning of 1977. The NMS cautions that the interpretation of such flows is not straightforward. Some registrants do not deregister on finding a job by another means and may remain on the register until deleted by NMS screening at the end of each quarter.

Walsh (1977A) has attempted to construct a time series of vacancies using the responses of firms to the CII-ESRI industrial survey, carried out since 1961. The series thus constructed was found to be negatively correlated with the registered unemployment rate. Walsh used the constructed series to form an estimate of the 'full employment' level of unemployment in Ireland. He concluded that even if half the existing firms in the manufacturing sector were reporting that a shortage of labour was a constraint on production, the national unemployment rate would not fall below 4.5%.

The labour force

The Trend of Employment and Unemployment contains annual estimates of the size of the total labour force, distinguishing the total at work by economic sector and the numbers out of work. Before the introduction of the Labour Force Survey in 1975, this was the only source of information on the size of the total labour force in Ireland. Unfortunately, the method by which these estimates were arrived at for non-census of population years has never been explicitly set out, and the figures have in the past been subject to substantial and frequent revision. The estimates contained in the Trend of Employment and Unemployment have now been superseded by those from the Labour Force Survey, classifying the total labour force by sector, and also by age, sex, and marital status. The Labour Force Survey is not without deficiencies. In particular, the results of the 1975 and 1977 surveys were subject to significant revision in the light of the results of the 1979 Census of Population.

Given an estimate of the underlying population, labour force participation rates may be calculated. Such participation rates are defined as the ratio of gainfully occupied persons to the total population in the relevant age, sex and marital status category. Detailed knowledge of such participation rates is of crucial importance in forecasting future trends in labour supply. The results of such forecasting exercises have been presented by Walsh (1975) and by Keating (1977).

Seasonal adjustment of employment and unemployment data
Analysis of short-term trends in employment and unemployment must take account of seasonal variation in these series. Successive monthly and quarterly figures will fluctuate due to seasonal variation in the demand for labour, independently of any underlying upward or downward trend in employment and unemployment. It is impractical to attempt to draw any conclusions about unemployment or employment trends without allowing for such variation.

Industries most affected by seasonal factors are the building and construction industry and those industries based on agriculture, such as food processing. In general, unemployment reaches its peak in January/February, then gradually declines to reach its lowest point about August/September, after which it begins to increase again. Seasonal variation in employment is the opposite of this pattern. For purposes of monitoring short-term trends in the economy it is essential to adjust monthly or quarterly employment and unemployment figures to eliminate such seasonal variation, giving the underlying seasonally corrected value. The method used by the CSO to seasonally correct its manpower series is the additive version X-11 of the US *Bureau of the Census Program* (Shiskin et al. (1965)).

The X-11 programme assumes that any realisation of a quarterly or monthly time series can be divided into three components: a trend cycle (C_t), a seasonal (S_t), and an irregular component (I_t). The relationship between these components may be

multiplicative $\quad Y_t = C_t \times S_t \times I_t$

or additive $\quad\quad Y_t = C_t + S_t + I_t$

where Y_t is the actual or 'observed' value of the series.

The programme produces various diagnostic statistics which can be used to decide which model is most appropriate to the series being analysed. Once the preferred model is specified the programme provides estimates of the magnitude of each component. The programme is an iterative one, successively refining its estimates of the three components. The seasonal component identified, called the seasonal factor, is then removed from the original series by subtraction (the additive model) or by division (the multiplicative model) to yield the 'seasonally adjusted' series.

Whether Irish unemployment statistics should be adjusted multiplicatively or additively has been the subject of considerable debate (see Dowling (1975), Bradley (1977) and O'Reilly and Gray (1980)). O'Reilly and Gray provide a concise explanation of the iterative process employed in the X-11 programme and also a summary description of an alternative programme which is used by the British Central Statistical Office which allows for the possibility that seasonality may be of a mixed nature characterised by both multiplicative and additive components. The authors conclude that an additive model is more appropriate for seasonally adjusting Irish unemployment data.

Appendix

The changing content of the Live Register over time

Statistics of unemployment are collected as a by-product of the administration of the Social Welfare Acts. This section lists the main changes in these Acts and their impact on registered unemployment.

January 1953   The Social Welfare Act, 1952 established a social insurance scheme for employees under which unemployment benefit and a number of other benefits are provided. Subject to certain minor exceptions, insurance was compulsory for all employed persons between the ages of sixteen and seventy earning less than £500 per annum.

January 1961   Family benefits extended to all children, instead of as previously only for the first two.

January 1965   The Social Welfare Act (Miscellaneous Provisions), 1965 extended unemployment assistance to some smallholders in certain western districts.

April 1965   Insurance exemption limit for non-manual workers set at £800 per annum.

Up to and including 1965, the total number of persons registered was known as the live register. The adequacy of this figure as a measure of the true level of unemployment was questionable, in particular in that many on the register could not properly be considered as unemployed. As from January 1966 the definition of the live register was changed by the exclusion of those classes of persons listed below who could not be considered as unemployed.

January 1966   The following categories of person were excluded from the definition of the live register with effect from this date: 1. persons who while being the beneficial owners of agricultural land are applicants for unemployment assistance (i.e. smallholders); 2. applicants for credits due to a trade dispute; 3. students at school or college and other persons signing on for temporary seasonal employment.

April 1966   Insurance exemption limit for non-manual workers raised to £1,200 per annum.

April 1967   Employment period orders abolished. Under these orders, men without dependents living in rural areas were excluded from the receipt of unemployment assistance between April and November of previous years. The orders reduced the live register by some 5,500 during their period of operation each year.

January 1968   The Redundancy Payments Scheme was introduced; qualified redundant workers received both a lump sum and a monthly payment, the duration of the monthly payment being dependent on length of service with their former employer.[15]

January 1968   The duration of unemployment benefit payments was increased from 156 to 312 days. This resulted in the inclusion on the live register of a number of smallholders who would previously have been excluded upon transferring to unemployment assistance when their title to benefit was exhausted.

October 1970   Payment of retirement pensions transferred from employment offices to post offices. The impact on the live register was estimated to be a reduction of 1,200 persons.

January 1971   Retirement pensions payable from the age of sixty-five. It is estimated that 1,600 persons were removed from the live register in 1972 as a result.

April 1971   Employment Period Orders reintroduced. From 14 April to 16 November 1971 men without dependents living in rural areas were excluded from the receipt of unemployment assistance. From 28 April 1971, however, this exclusion ceased to apply to such men residing on specified coastal islands who were fifty or more years of age. Men who were affected by the Employment Period Order and who continued to register for credits or for work were included in the live register in the category 'others'.

July 1973   The age-limit for eligibility for old age pensions was reduced from seventy to sixty-nine. This resulted in a decline of 800 persons on the live register.

April 1974   Income limit for insurability abolished. All excluded employees henceforth compulsorily insured.

April 1974   Pay related unemployed benefit introduced — payable for up to 147 days.

July 1974   Reduction in the age-limit for old age pensions to sixty-eight.

April 1975   Reduction in the age-limit for old age pensions to sixty-seven. About 700 people were removed from the live register as a result.

April 1976   Extension of the duration of unemployment benefit payments from 312 to 390 days. Pay related benefit payable for up to 381 days.

October 1977 Reduction in the age-limit for old age pensions to sixty-six.

October 1978 Restrictions on the entitlement of single women and widows to unemployment assistance relaxed. It is estimated that the relaxation of this restriction added some 2,300 to the live register initially, falling to 2,100 by end March 1979.[16]

January 1980 Persons on systematic short-time working and persons aged sixty-five and over excluded from the live register. This reduced the live register by about 3,800.

Notes to Chapter 2

[1] Although, of course, geographical coverage changed after the 1911 Census.
[2] For a detailed description of the Labour Force Survey and revised estimates for 1975 and 1977 see: Labour Force Survey 1979 Results, PL113, Stationery Office, Dublin.
[3] Data from the first three surveys are compared in the report of the 1979 survey.
[4] See *Irish Statistical Bulletin* December 1977, pp. 267-77.
[5] See *Irish Statistical Bulletin* September 1979, pp. 261-8.
[6] Available from the Research and Planning Division, AnCO.
[7] See Neary (1975) for a detailed description of the Survey.
[8] The responses to this question were used by Walsh (1977A) to construct a series on vacancies.

[9]   See *Irish Statistical Bulletin* March 1973, pp. 18-19.

[10] Available from the National Manpower Service, Dublin.

[11] See *Irish Statistical Bulletin* March 1980, pp. 13-26.

[12] Most current statistical series, such as those based on the live register, are published speedily in the form of special releases by the CSO, and normally appear in this form several months prior to publication in the *Statistical Bulletin* or the *Trend of Employment and Unemployment*.

[13] Walsh's (1978) analysis concentrates solely on the supply side of the labour market. The increase in employers' social insurance contributions, levied in order to finance enhanced benefits, may have contributed to increased unemployment by inducing a decline in firms' desired level of employment (Kirwan (1979)).

[14] Published in the *Trend of Employment and Unemployment* and in the *Manpower Information Quarterly*.

[15] See Whelan and Walsh (1977) for an analysis of the impact of this scheme on the job search behaviour of the unemployed.

[16] For details of the initial monthly impact of this measure on the live register total see *Manpower Information Quarterly* Vol 1, No 1, Table 34.

3 Agriculture

The agricultural sector (including fishing and forestry) accounts for about 12-13 per cent of national income, and employs directly around 20 per cent of the labour force.[1] Though the share of agriculture in the national income has declined continuously for many years, as have the numbers engaged in it, it is scarcely necessary to emphasise its importance to the Irish economy. While the share of industry in the national income now exceeds that of agriculture, a considerable proportion of industrial production and employment arises in food processing, so that much of the nation's economic activity is based, directly or indirectly, on agriculture.[2]

There is a considerable variety of data published in relation to agriculture in Ireland. This chapter does not attempt to cover all of these sources, but only those regularly published series relating to land utilisation, crops and livestock, employment, gross and net output, incomes and prices, including index numbers of output and prices. These data are published regularly in the quarterly *Irish Statistical Bulletin* and later in the annual *Statistical Abstract*. They include the results of the annual June enumeration, the winter livestock enumeration, the April and August pig enumerations, the annual estimates of agricultural output and data collected from agricultural markets. Many of these series are very comprehensive, and are useful for a variety of detailed analyses including trends in the structure and composition of output, employment, land utilisation and size of holdings.

There are also a considerable number of special surveys and reports relating to particular aspects of agriculture. A number of these special publications are referred to below. Sources of statistics of fishing and forestry are also briefly discussed.

Agricultural output

Two of the most important and comprehensive series of agricultural statistics are the results of the annual agricultural enumeration, taken in June of each year, and the annual estimates of agricultural output.

The annual series of June enumerations extends back as far as 1847, with occasional interruptions. Up to and including 1953 a complete enumeration of holdings was undertaken, but the current practice is to carry out a complete enumeration (census) at intervals of five years, and to rely on sample inquiries for intervening years. The use of sample enumerations (which usually cover about half of the area of the country) subjects the results to sampling errors (in addition to any other types of error) and revisions to the raised estimates may be necessary when the results of the following census become available. The principal aim of the enumeration is to determine the total area under various types of crops, total livestock numbers, the number of males engaged in agriculture and the numbers and types of agricultural machinery in use. Various special inquiries are also included from time to time.

For years in which there is a complete enumeration, results are published at rural district,[3] county, provincial and national level. In non-census years estimated results for the main items are published at county, provincial and national level. Because of the time-lag in the production of detailed results from the enumerations, provisional national estimates are produced on the basis of a subsample covering about a quarter of the area of the country, normally published in the September issue of the Irish Statistical Bulletin. The amount of information published has tended to vary from year to year.

Provisional estimates of the area under crops and livestock numbers in June of each year are usually published in the September issue of the Statistical Bulletin, under the rubric 'Agricultural Statistics'. These record (for the whole country) estimates of the area under certain crops (wheat, oats, barley and potatoes) and numbers of cattle and sheep, for the current year and the previous year. For some years, more comprehensive analyses of the areas under crops and livestock numbers, including breakdowns by county and province, are published in later issues of the Bulletin, but there has been a tendency recently to reduce the volume of published data for non-census years.

The results of the June enumeration provide a useful 'cross-sectional' analysis of the structure of agriculture from year to year. These statistics have been collected for many years and are therefore a valuable source for analyses of long and short term trends in Irish agriculture.

Other annual enumerations are the winter livestock enumeration, the April and August pig enumerations and the EEC-wide *Farm Structure Surveys*. The first winter livestock enumeration was taken in January 1934, and since 1938 has been taken annually (except for 1944 and 1946). From 1972, the enumeration has been taken on 1 December instead of 1 January, to ensure greater harmonisation with the corresponding statistics of other EEC member states. The results, usually published in the following March edition of the Statistical Bulletin, record estimates of the numbers of various categories of cattle, sheep, pigs and poultry in the country at the beginning of December, with the corresponding figures for the previous year and the actual and percentage changes over the year. Livestock numbers vary quite considerably during the year, and the winter enumeration is taken at a time when numbers are usually at a minimum, due to the slaughter and export of livestock at the end of the grazing season and before the birth of new stock. As the June enumeration is taken at a time when livestock numbers are at, or near, a maximum, the results of both enumerations are of interest in recording the range of the livestock cycle from year to year.

The winter enumeration is also based upon a sample, covering approximately a quarter of the area of the country, and 'blown up' to provide estimates for the whole country.

Analyses of agricultural holdings by size of cattle and pig herds, based on the December enumeration, have been carried out every two years from December 1973 and published in the Statistical Bulletin. These include frequency distributions of the number (and percentage) of holdings and of animals classified by herd size; recent results show a trend towards larger herd sizes concentrated in a smaller number of holdings, reflecting the pressures towards economies of scale in the livestock sector. To date these analyses have been carried out for cattle and pigs only.

Two further sample enumerations have been held annually since 1973 in April and August in respect of pig stocks. The samples cover approximately 35 per cent of total pigs and the results are published in the June and September issues of the Irish Statistical Bulletin.

A further development has been Ireland's participation in the EEC-wide Farm Structure Surveys, the first of which were carried out in 1966/67 and 1969/70, before Ireland acceded to membership. These

are aimed at obtaining EEC-wide harmonised and comparable
statistics for use in connection with the various farm structural policies.
The first survey in which Ireland participated as an EEC member was
in 1975 and since then there have been further surveys in 1977 and
1980. Surveys are also planned for 1983 and 1985.

The EEC surveys exceed the range of data traditionally collected in
the annual June enumerations and have as their basic unit of inquiry
the farm (i.e. area worked) as distinct from the holding (i.e. area
owned) which is the unit used in Irish statistics. The 1975 FSS, for
example, involved the collection of data on an extensive range of crops,
the reintroduction, after many years, of rough grazing as a separate
item for enumeration, the collection of labour input (as distinct from
labour force) statistics and the collection of data on ages of holders,
members of their families and hired workers, use of machinery (own
and hired), contract sales, managers' agricultural education, etc.
While the national agricultural census forms include as many as
possible of the EEC data demands, the need to collect considerably
more information coupled with the problem of compiling farm-based
statistics, has resulted in supplementary sample inquiries (some 30,000
to 40,000 holdings) being carried out to supplement the basic
enumeration information and to provide the additional data required
to convert holdings to farms. The 1975 survey results have been
published in considerable detail by the EEC Statistical Office and some
1977 results are also available from the same source.

In the broader context of the role of the agricultural sector in the
economy of the country, and of the contribution of that sector to the
total annual output of goods and services, the annual estimates of
agricultural output are the most important aspect of agricultural
statistics.

Estimates of agricultural output are published each year in the
Statistical Bulletin. Output estimates for the Republic of Ireland were
first made for 1926-27, then for 1929-30, and annually since 1934-35.[4]
Up to 1944, the estimates related to 'agricultural years' rather than
calendar years — for livestock the agricultural year ended on 31 May,
and for most crops on 30 September — but the estimates now relate
more conveniently to the calendar year.

Estimates for individual items of output are based upon a variety of
sources, including the June enumeration, the *Census of Industrial
Production* (for data relating to purchases of agricultural produce), the
external trade statistics, reports from markets, special sample

inquiries, and data collected for administrative purposes by the Department of Agriculture and the various agricultural promotion and marketing bodies. Notes on the methods of estimation for constituent items are included in the reports for earlier years, published in the Statistical Bulletin, and in certain of the later reports, particularly in years when there have been changes or revisions in methods of estimation.

The annual estimates are published (usually in the June issue of the Statistical Bulletin) in the form of a report with text and tables: the text includes definitions of certain terms used in the report, notes on some of the estimates, and a commentary on the results. It will be useful to consider some of the terms used in the report.

Gross agricultural output is that part of total agricultural production sold off farms plus the estimated value of agricultural produce consumed by farm households, plus the estimated value of changes in livestock numbers. It may be (and sometimes is) defined to *exclude* the value of changes in livestock numbers. Both concepts of gross output have their particular uses and limitations.

It is important to appreciate the distinction between gross output as defined above and total agricultural production. The definition and treatment of constituent items of output are explained in the text of the annual report and these notes should be read carefully before using the statistics. As explained, agricultural output includes that part of agricultural production sold off farms to other sectors (plus estimated consumption by farm households). Produce which is retained on farms (for example, as seed or animal food), or which is sold by one farmer to another, is not included in output. In addition, the values of certain crops, particularly oats and potatoes for seed and whole oats for feed, which are sold off farms and then repurchased by farmers as inputs, are also excluded from output. Only the amounts of these crops which are sold permanently off farms are included with output.[5] In the case of other crops such as barley, wheat and grass seed all sales off farms are included as part of output, while amounts of these products repurchased by farmers are included, at retail prices, as inputs.

The estimates of the value of gross output are based on the prices which farmers are estimated to receive for their produce. Output estimates are derived from a variety of sources, and a number of adjustments are made to account for transport costs and other marketing margins which may be included in the prices of agricultural produce at their point of valuation (for example at factory, port or market). As far as is

Table 3.1 Estimated output of agricultural products, 1977-1979

Product	Unit of quantity	Estimated quantity			Estimated value (£000)		
		1977	1978*	1979	1977	1978*	1979
Livestock:							
Horses	No (000)	15	11	10	12,107	11,547	10,524
Cattle & Calves§	,,	1,768	1,903	1,629	517,611	613,415	573,883
Sheep and Lambs	,,	1,350	1,604	1,391	32,173	52,824	54,318
Pigs	,,	2,028	2,108	2,314	107,866	121,075	135,652
Poultry:							
Turkeys	,,	1,154	1,232	1,370	7,692	9,115	10,687
Geese	,,	60	61	69	437	519	617
Ducks	,,	1,158	1,123	1,158	1,737	2,190	2,432
Ordinary fowl (incl. value of day old chicks)	,,	24,932	25,045	26,563	20,704	20,763	26,656
Total Livestock	value	—	—	—	700,327	831,448	814,769
Livestock Products:							
Milk: a. consumed by persons	litres (m)	632	631	632	65,169	71,489	75,244
b. used in industry	,,	3,384	3,911	3,993	351,835	443,690	465,350
Wool	tonnes	6,266	6,088	6,129	5,608	5,811	5,742
Hen eggs	million	605	598	566	19,658	18,043	20,783
Other Livestock Products	value	—	—	—	2,419	2,506	2,124
Total Livestock Products	value	—	—	—	444,689	541,539	569,243
Total Livestock and Livestock Products	value	—	—	—	1,145,016	1,372,987	1,384,012
Crops:							
Wheat	tonnes (000)	244	246	242	22,022	23,344	22,815
Oats	,,	29	36	42	2,439	3,293	3,834

Table 3.1 continued. Estimated output of agricultural products, 1977-1979

Product	Unit of quantity	Estimated quantity			Estimated value (£000)		
		1977	1978*	1979	1977	1978*	1979
Barley	,,	1,083	1,129	1,121	95,249	96,669	99,551
Turnips	,,	40	40	42	377	551	809
Sugar Beet	,,	1,376	1,456	1,322	29,500	35,196	33,767
Potatoes	,,	401	389	336	26,992	15,785	41,789
Cabbage	,,	97	106	107	9,723	11,609	14,404
Hay	,,	19	18	19	758	808	1,032
Straw	,,	36	38	40	363	505	577
Fruit	value	—	—	—	5,969	6,410	5,634
Other crops (incl. grass seed)	,,	—	—	—	28,587	28,073	29,742
Total Crops	value	—	—	—	221,979	222,243	253,954
Turf	tonnes (000)	1,069	994	940	11,304	11,491	13,595
Total Crops & Turf	value	—	—	—	233,283	233,734	267,549
Gross value of agricultural output (excl. value of changes in livestock nos.)					1,378,299	1,606,721	1,651,581
Value of changes in livestock nos.					-12,442	-13,664	+25,868
Gross value of agricultural output (incl. value of changes in livestock nos.)					1,365,857	1,593,057	1,677,429
Farm materials purchased by farmers: Feeding stuffs					225,372	262,129	348,495
Fertilisers (including lime)					100,335	135,634	158,027
Seeds (including cleaning and retailing charges)					15,932	17,622	20,210
Total value of farm materials					341,639	415,385	526,732
Net Value of Agricultural Output (incl. value of changes in livestock nos.)					1,024,218	1,177,672	1,150,697

*Revised. §The value of cattle output includes payments under the Bovine Tuberculosis and Brucellosis Eradication Schemes.

Source: ISB, June 1980

practicable, output is valued net of distributive margins. Agricultural produce consumed on farms by farmers and their families is assumed to be identical in quality with produce sold off farms and (except in the case of milk) is valued accordingly.

The *net value of agricultural output* or *net agricultural output* is defined as gross agricultural output less the value of certain farm materials used as inputs. Since gross agricultural output may be taken to include or exclude the value of changes in livestock numbers, net agricultural output may also be estimated on either basis.

The value of certain farm materials, deducted from gross output to arrive at net output, includes the estimated cost of fertilisers, feeding stuffs and seeds purchased by farmers and used for current production. The estimates recorded under this heading, however, do not measure total outlay by farmers on these inputs. The value of 'own produce' used as inputs by farmers, as well as the value of produce purchased as inputs by one farmer from another, are excluded from inputs. The value of sales of potatoes and oats used for seed, and oats used for feed, are excluded from output; the value of purchases of these products by farmers is similarly excluded from inputs. Only the estimated retail and cleaning charges of purchased Irish seed potatoes and oats are included as inputs, plus the cost of any imports of these products. The figures for farm materials purchased as inputs, published in the annual estimates, understate the actual value of agricultural inputs. The value of net agricultural output, however, is unaffected by this treatment, since the value of these transactions is excluded from both the output and input sides of the accounts.

The estimates for net agricultural output should not be interpreted as measuring income arising, or generated, in the agricultural sector. Several adjustments must be made to agricultural net output to arrive at agricultural incomes. These are explained in Table 3.2.

The principal results of the estimates of agricultural output for 1979 are recorded in Table 3.1. Details of the quantity and value of output of the constituent items should be interpreted in relation to the definitions of gross agricultural output given above, i.e. they are estimates of sales off farms plus agricultural produce consumed by farm households.

Other tables in the report analyse in greater detail certain aspects of agricultural output — changes in the volume of output of principal commodities, the output of cattle and calves, total milk production and disposal, and estimates of the value of farm produce and fuel consumed on farms without process of sale. Another table records estimates of the value of exports of various kinds of agricultural produce; these include not only the value at agricultural prices of 'direct' agricultural exports but also estimates of the agricultural content of other exports — such as tinned meat, butter, textiles, ale, whisky, and confectionery.

The relation between total agricultural production and gross agricultural output has been discussed above, and it has also been explained — and illustrated in Table 3.1 — how the estimates of net agricultural output are obtained. However, not all costs incurred in producing gross output are included in the value of farm materials recorded in Table 3.1, and to this extent net agricultural output is not equivalent to incomes arising in agriculture. Certain other expenses must be deducted from net output to arrive at estimated agricultural income. Details of these expenses for 1979 are recorded in Table 3.2; they include rates, repairs to machinery, estimates of depreciation of implements and machinery, petrol and fuel oil, and certain marketing expenses.[6] The deduction of these expenses — non-factor inputs — from net agricultural output, plus subsidies not related to sales, gives total factor income arising in agriculture.

Total agricultural income is distributed as payments of Land Annuities, wages and salaries and — as a residual — income from self-employment and other trading income. The residual figure of income from self-employment, which represents the estimated income of farmers and their families, includes a significant non-monetary element — the estimated value of farm produce consumed by farm households, and the value of changes in livestock numbers. Moreover, this estimate is a compound of labour remuneration, profit and interest on capital employed.

These estimates of agricultural income are important as a measure of the share of the agricultural sector in total national income and of the contribution of that sector to the total annual output of goods and services. As such the estimates are included in the annual national income and expenditure accounts, discussed in Chapter 6.

The estimates of gross and net agricultural output in Table 3.1 are recorded at current money values, i.e. the quantities of output of the constituent items are valued at current prices. Thus changes in the value of gross and net output are a compound of changes in prices and changes in quantities produced. It is naturally of interest to examine the independent effects of changes in prices and changes in quantities on the value of agricultural output. The principal means by which this is done is by the use of index numbers of the volume of output and indexes

Table 3.2 Income arising in Agriculture, 1977-1979 in £ million

	1977*	1978*	1979
Net output of agriculture (inc. value of changes in livestock numbers)	1,024.2	1,177.7	1,151
Expenses of agriculture:			
Rates	23.4	33.6	39
Repairs to machinery, spare parts etc.	24.7	32.8	
Petrol, oil, etc.	37.5	41.1	
Depreciation of machinery and implements	88.8	112.4	344
Transport and marketing	19.0	21.8	
Other	60.0	73.4	
Total expenses	253.3	315.1	383
Net output less expenses	770.9	862.6	768
plus subsidy§ under Land Acts	1.9	2.0	2
plus subsidies not related to sales (livestock headage payments in handicapped areas, beef cow subsidy scheme, etc)	21.3	24.6	26
Income arising in agriculture of which:	794.1	889.2	796
Land Annuities (rent element plus subsidy included above)	3.0	3.0	3
Wages and salaries (including employers' contribution to Social Security)	47.0	50.0	54
Income from self-employment and other trading income	744.1	836.2	739

*Revised.
§Interest element of Exchequer payment to Land Bond Fund.

Source: Irish Statistical Bulletin, June 1980

of agricultural prices; index numbers of the volume of output are published in the annual report of the Minister for Agriculture, and these are discussed below. Index numbers of agricultural prices are discussed in Chapter 8 (Wages and Prices).

Statistics relating to certain aspects of agricultural output are also recorded in other series published in the Statistical Bulletin. The output estimates described in Table 3.1 relate only to sales off farms plus consumption by farm households. Estimates of the *total production* of certain crops (wheat, oats, potatoes, sugar beet, etc.), as well as the estimated yield per statute acre, are also published each year in the Statistical Bulletin.

These estimates are subject to some margin of error, and the CSO advises that 'the estimates . . . should not be regarded as absolute levels of yield but rather as indicating trends'. The estimates for the area under crops are based upon the June enumeration, and the figures for total yield are obtained by applying to these estimates figures for yield per statute acre. (In the case of sugar beet, however, the area under the crop is divided into total production to give the estimated yield per statute acre). These figures relate to the estimated total production of crops, including amounts of those crops which are used as feeding stuffs or seeds on farms, and therefore differ from the *output* estimates recorded in Table 3.1.

Particulars relating to agricultural output are also recorded in the 'Economic Series' which appear in each issue of the Statistical Bulletin. These series record monthly data for the current year and the preceding three years; each series is graphically illustrated and provides a useful picture of short-term trends. Series relating to agricultural output record exports of cattle (number) and pigs received at bacon factories (number). Other series indirectly related to agricultural production record exports of fresh and frozen beef, creamery butter production and wheaten flour production. These series demonstrate seasonal variations in the production of most agricultural commodities; creamery butter production, for example, reaches a seasonal peak in May-June and declines to a seasonal minimum in December-January.

Statistics of the stock and sale of pigs are published regularly in the Statistical Bulletin. Data on pig stocks are collected in the April, June, August and December enumerations (discussed above). Monthly returns of pigs (number and weight) purchased by bacon factories are also compiled and published in the Bulletin. Based on these and other sources, a brief survey of the pig industry is compiled and published annually in the Bulletin. Additionally, an analysis of the number of holdings with pig herds is carried out every two years.

Employment and earnings

Regular statistics of employment and earnings in agriculture derive from two main sources. Information on males only, working on farms is collected in the annual June enumeration of holdings; results are published in the *Statistical Bulletin*. These record the number of males engaged in farm work, distinguishing members of family and 'Other Males', both permanent and temporary. Separate figures are given for males aged fourteen to seventeen and eighteen years and over. A second table disaggregates the national data by county and province.

Since 1975 the CSO has carried out a regular survey of the earnings of permanent agricultural workers. The sampling frame for this survey is the list of holdings in the annual June enumeration reporting one or more permanent male employees, and approximately 1,800 holdings are presently included in the survey. Data are collected on the age, sex, gross earnings, benefits and hours worked for a selected reference week, for all permanent workers, defined as 'non-family workers, male or female, who are employed permanently, i.e. liable to be continuously employed on a full-time regular basis throughout the year.' Regular part-time, seasonal and occasional workers are excluded. While data on permanently employed female workers is collected in the survey, the numbers involved are small and estimates of average earnings and hours worked are not reported in the results of the survey, on the grounds that potential sampling errors are too great.

Principal results of the survey, published annually in the Statistical Bulletin, report average earnings and paid hours of work for permanent male workers, classified by occupational status (managers/foremen, stockmen, horticulturists, etc), by type of work in which normally engaged (livestock, horticulture, mixed), by age group, and by number of permanent workers employed on holding. Earnings are estimated with and without 'benefits' i.e. the actual or imputed value of accommodation and/or meals which may be provided by the employer free or in lieu of payment. Between 1975 and 1980 this survey was conducted annually, but there was no survey in 1981 and it is intended that in future the survey should be conducted biennially.

The aforementioned sources are specifically directed to analysis of agricultural employment. Details of agricultural employment also come from the biennial *Labour Force Survey* (see Chapter 2), including estimates of agricultural employment by sex and planning region, by sex and age-group, and by employment status.

Index numbers of the volume of agricultural output

The annual estimates of agricultural output and costs of farm materials are valued at current agricultural prices; changes from year to year in the value of output may be due to changes in prices, or to changes in quantities, or more generally to a combination of both. If prices remained constant from year to year, changes in the value of agricultural output could be attributed solely to changes in quantities. On this assumption, suppose we now express the value of output in each year as a percentage of the value of output in some selected 'base year'. It does not matter which year is selected as the base; output in each year is related to the value of output in the fixed base year. Expressed as percentages, output in the base year will take the value of 100, while the value of output in every other year will be related to this base year value. As an example, suppose the value of output in year 1 is £1,000m, in year 2 is £950m, in year 3 is £1,090m and in year 4 is £1,195m. Taking year 1 as base year, we now express the value of output in each year as a percentage of the base year output of £1,000m. This gives

$$\frac{1,000}{1,000} \times 100 = 100 \text{ for year 1}; \quad \frac{950}{1,000} \times 100 = 95 \text{ for year 2}; \text{ and}$$

similarly values of 109 and 119.5 respectively for years 3 and 4. The series for agricultural output is now said to be expressed in index number form; the index for year 2 is 95, the index for year 3 is 109 and so on.

The absolute (though not the relative) value of the index for any year depends upon the base year selected, and this should be remembered in interpreting the index. In the example above, output in year 2 is 5 per cent lower than in year 1; output in year 3 is 9 per cent higher than in year 1 and output in year 4 is 19.5 per cent higher than in year 1. If a different year is selected as base year, the absolute values of the indexes will be different — e.g. with year 2 as base year, the above series becomes 105.3, 100, 114.7 and 125.8 for years 1 to 4 respectively — but the relative values are unchanged.

It was assumed above that prices remained constant so that changes in the value of output arose only through changes in the volume of output. In this case the series of indexes calculated may be regarded as index numbers of the *volume of agricultural output*. In practice, changes in the value of output will be a compound of price and volume changes; the index number series derived from these current value aggregates will be index numbers of the value of agricultural output. To calculate indices of the volume of output it is first necessary to eliminate the effects of price changes by re-valuing the quantities of output in each year at constant prices, giving a series of the value of agricultural output at constant prices. These constant price aggregates may then be expressed in index number form.

The total value of output in any one year is obtained by valuing the quantities produced at the average prices of that year. If, however, the quantities produced in each year are valued at *constant* prices — say 1975 prices — we obtain a series for agricultural output at constant 1975 prices; this is then converted to index number form to give index numbers of the volume of agricultural output. By this means changes in the volume of output may be measured. (For a more extensive discussion of volume index numbers see Chapter 4.)

Index numbers of the volume of agricultural output in Ireland for 1975-79 are recorded in Table 3.3. Index numbers are calculated not only for total gross output (excluding and including changes in livestock numbers), but also for the major constituents of output — crops, crops and turf, and livestock. Also recorded are indexes of net agricultural output and inputs of farm materials — the calculation and interpretation of these indexes will be explained below.

The base year for each index number series is 1975 = 100. Output in each year is re-valued at constant 1975 prices, and expressed as a percentage of the value of output in 1975. Thus the volume of gross agricultural output in 1979 (including changes in livestock numbers) was 10.4 per cent higher than the volume of output in 1975. The constant prices used to re-value output in each year are 1975 prices, the base year prices. It is not necessary that the constant prices used to re-value output should be those of the base year for the series, although it is more convenient; any set of constant prices would suffice, as long as the same prices are used for each year.

The last row of Table 3.3 records index numbers of the volume of farm materials used as inputs, calculated in the same way as the index numbers of volume of gross output; quantities of farm materials purchased in each year are re-valued at constant 1975 prices, and the resulting constant price aggregates are expressed as an index number series (base year 1975 = 100). Between 1975 and 1979 the volume of farm materials used as inputs rose by 75.2 per cent.

The remaining index numbers in Table 3.3 are index numbers of the volume of net agricultural output. We shall first explain how these indexes are calculated and then briefly discuss the concept of volume of net agricultural output. The estimates of net agricultural output at current prices in Table 3.1 are obtained as the difference between gross agricultural output and farm materials purchased as inputs. For calculating volume indexes, gross output and farm materials are re-valued at constant prices. Estimates of the value of net agricultural output at constant prices are obtained as the difference between gross output at constant prices and the value of farm materials at constant prices. The series for net output at constant prices is then converted into index number form, to give

Table 3.3 Index numbers of the volume of agricultural output
(1975 = 100)

	1976	1977	1978*	1979
Crops	96.0	120.8	123.8	118.8
Crops and Turf	96.2	119.2	121.5	116.6
Livestock Products	107.5	115.7	129.2	131.4
Excluding value of changes in livestock numbers:				
Livestock	76.7	86.9	90.0	85.4
Livestock & Livestock Products	86.9	96.5	103.0	100.7
Gross Output	88.4	100.0	105.9	103.2
Net Output	80.9	93.1	94.2	82.8
Including value of changes in livestock numbers:				
Livestock	89.0	94.3	98.6	97.0
Livestock & Livestock Products	95.6	101.9	109.5	109.2
Gross Output	95.7	104.7	111.5	110.4
Net Output	89.8	98.7	100.6	90.8
Farm Materials	115.0	124.5	147.5	175.2

*Revised
Source: ISB, June 1980

the volume index series recorded in Table 3.3. Thus, to calculate the index of volume of net output in 1979, we first subtract the value of farm materials in 1979, valued at 1975 prices, from the value of gross output in 1979, also valued at 1975 prices — this gives the value of net agricultural output at constant (1975) prices. The latter is then expressed as a percentage of the value of net output in 1975.

This method[7] of calculating volume indexes of net output is different from that used to measure the volume of gross output or farm materials. The value of gross output is directly calculated by re-valuing quantities at constant prices: this can be written

$$\sum_{i=1}^{n} p_i q_i,$$

where $q_1, q_2, \ldots q_n$ are the constituent items of output and $p_1, p_2 \ldots p_n$ are the (constant) prices used to value these commodities. The value of farm materials at constant prices is similarly calculated. Net output, however, does not correspond to any simple price × quantity relationship — it is derived as the difference between two other aggregates, and does not lend itself to direct calculation.

Index numbers of the volume of gross and net agricultural output and farm materials are extremely useful. By eliminating the effects of price changes, they stand as measures of the rate of growth of output (and input) in 'real terms', and may be used to compare rates of growth in different sectors of the economy. In conjunction with other statistics, such as numbers employed in agriculture, they may also be used — more hazardously — to indicate trends in productivity.

The present index number series replaced a series with base year 1968 = 100, which in turn had replaced a series based on 1953 = 100. The earlier series are reproduced in the annual Statistical Abstract.

Corresponding to the series on the volume of agricultural output and input are regular series of statistics and index numbers of agricultural prices. These are described in Chapter 8.

Other sources

The series described in the preceding sections — the June and December enumerations, estimates of agricultural output, employment and earnings and index numbers of agricultural output — are the most important of the regularly published statistics (along with statistics of agricultural prices, discussed in Chapter 8). All these series are published in the Statistical Bulletin and later reproduced, wholly or in part, in the Statistical Abstract. A complementary source of data is the annual report of the Minister for Agriculture which includes tables of agricultural output, prices, exports, etc as part of its review of agricultural progress and policy.

A valuable source of information on the structure and performance of agricultural holdings is the annual *Farm Management Survey* carried out by An Foras Talúntais. This comprises a sample of 1,300-1,500 farms of five or more acres, stratified by size of holding (six categories) and county (26).[8] Its primary purpose is to collect and analyse data on farm output, costs and incomes and their variation between different types and sizes of farms and different areas of the country. The published reports contain a wealth of information on the distribution and structure of farming (by size, soil type and system of farming), on output, labour inputs and farm incomes, and on costs of production. Numerous cross-classifications are provided and the computerised data base permits a wide range of analyses.

Agricultural statistics for Ireland and the other member countries are published regularly by the Statistical Office of the European Communities. The 1975 *Community Survey on the Structure of Agricultural Holdings* includes detailed analyses of the structure of agriculture in each of the member states. Data are provided by the statistical offices of the member states — the Irish survey and its relationship to the June enumeration of 1975 was discussed above. While deriving from the same source as the June enumeration, the Community survey presents different tabulations and forms of classification.[9] Characteristics examined include size, tenure, land use, labour, machinery and equipment, woodlands, crops and livestock.

Fishing and forestry

For fisheries, the most comprehensive source of data is the annual report on Sea and Inland Fisheries, formerly published by the Fisheries Division of the Department of Agriculture and Fisheries, and now by the Department of Fisheries and Forestry. It contains detailed information on the quantity and value of all fish caught and landed at Irish ports, and other data on prices, exports and imports and employment. Some statistics are also published in the annual Statistical Abstract, and in the annual reports of Bord Iascaigh Mhara. Statistics of the quantity and value of different categories of fish landed at Irish ports are also published in the Statistical Bulletin. The September issue records the quantity and value of fish landed by Irish-registered vessels in the period January-June, with corresponding figures for the previous year. Data for the twelve-month period January-December is reported in the following March issue of the Bulletin.

From 1933 to 1977 forestry was the responsibility of the Department of Lands (Forestry Division), but it is now incorporated with Fisheries in a new department. Detailed statistics of forestry operations are published in an annual report which includes data on land acquisition, the area under forestry (by county and region), planting operations, employment, income and expenditure, privately-owned forestry schemes and other relevant material. Some of these data are also recorded in the Statistical Abstract.

Quarterly statistics of forestry operations are published in the Statistical Bulletin, including the area of land acquired for planting purposes in the previous quarter, the number of men employed in the last week of each month of the quarter, and wages paid over the period. The statistics of land acquisition and employment are broken down by county, with regional sub-totals.

Notes to Chapter 3

[1]   Central Statistics Office.
[2]   In 1976, for example, food and drink establishments accounted for about 35 per cent of the total net output of all manufacturing establishments covered by the Census of Industrial Production.
[3]   There are 158 rural districts.
[4]   Earlier enquiries, relating to the whole of Ireland, were conducted in 1908 and 1912-13. See *The Agricultural output of Ireland 1908* (1912). The results of the 1912 inquiry were not published as a report due to the outbreak of the 1914-18 war; see however the *Report of the Departmental committee on food production* (Cd 8158, 1916). The estimates for 1926-27 were published as *The Agricultural output of Saorstát Eireann 1926-27* (P. No 132) by the Statistics Branch of the Department of Industry and Commerce. The estimates for 1929-30 were not published as a separate report, but results of the inquiry were published in later editions of the *Statistical Abstract*. Since the inquiry of 1934-35 the estimates have been published annually in the *Statistical Bulletin*.
[5]   However, oats sold off farms which are later repurchased by farmers in a processed form (for feed) are included as part of output.
[6]   Also included as expenses are miscellaneous Government and EEC levies paid by farmers.
[7]   This method of calculating net volume of output, called the 'double deflation' method, was developed by Dr. R. C. Geary, former Director of the Central Statistics Office. See R. C. Geary: 'The concept of net volume of output with special reference to Irish data' *Journal of the Royal Statistical Society* Vol. CVII, Parts III-IV, 1944.
[8]   Details of the sampling procedure are given in the published reports of the survey. Results based on a calendar year are available from 1972. Earlier surveys based on financial years were carried out by An Foras Talúntais for 1966-67, 1967-68 and 1968-69.
[9]   *Community Survey on the Structure of Agricultural Holdings 1975* (6 Vols), Statistical Office of the European Communities, Luxembourg, 1979.

4 Industrial Production

'Industrial production' includes three broad categories of economic activity: non-agricultural extractive industries (mining, quarrying, turf production), manufacturing industries, and certain service-type industries (e.g. public utilities).

The industrial sector has grown rapidly in the last two decades, as the country has sought to reduce its dependence on agriculture and to provide greater employment opportunities. In 1978, the sector 'industry' accounted for 34 per cent of gross domestic product at factor cost. The comparative experience of other developed countries suggests that the share of industry in GDP is unlikely to exceed this level in the future and in fact may decline, while services (the tertiary sector) will contribute an increasing share.

Historical development

Almost all sources of statistics of industrial production in Ireland date from 1926, when the first census of industrial production was carried out. Before this date, a certain amount of information on industrial production in Ireland was collected as part of the UK censuses of production, taken in 1907 and 1912 by the Board of Trade. Separate figures for the whole of Ireland were published in the final report of the 1907 Census,[1] but are of little use for comparative purposes in relation to subsequent census statistics.[2] The results of the second census (1912) were never published due to the outbreak of the 1914-18 War, but some of the census material was later provided by the British Board of Trade to the Department of Industry and Commerce, and was included in the *Report on the Census of Industrial Production,* 1926 and 1929.[3] Figures were published for gross and net output, materials used and numbers employed, for thirty-one industries, but again are not comparable with those for later years. They are, however, of interest in reflecting (approximately) the relative importance of different industries at that time.

The first census relating to Ireland (26 counties) alone, in 1926, was followed by a second in 1929, and from 1931 the census has been annual. The *Census of Industrial Production* is the most important source of information on industrial production in Ireland, and the rest of this chapter is concerned with the information collected in the census and the presentation and interpretation of census data. Other sources of data on industrial production are also discussed.

Since 1926, there have been changes in the scope and coverage of the census, and in the industrial classification used in the compilation and analysis of the statistics. The present framework of the census and the related statistical series date from 1973, when the industrial classification used in the census was altered to accord with the general classification of economic activities used within the European Community, known as NACE (Nomenclature Générale des Activités Économiques dans les Communautés Européennes). Previously (i.e. between 1954 and 1973) the census classification scheme followed the UN International Standard Industrial Classification (ISIC). The NACE and ISIC systems are not readily compatible, so that detailed industry-level series of output, employment and so on for years before 1973 are not comparable with the new series. Results of the 1973 census were compiled and published on both the old (ISIC) and new (NACE) bases, so that 1973 is the 'link year' between the two systems. Even at the aggregate level for all industries, there are differences in output and employment resulting from differences in coverage between the two systems.[4]

The Census of Industrial Production

The results of the annual census are published in the *Irish Statistical Bulletin* at various dates subsequent to the census. Individual industry reports are published when results become available, and the final summary results for all census industries are usually published within a few years of the census year concerned.

Preliminary reports published in various issues of the ISB subsequent to the census year summarise the detailed census results for individual industries. Under the NACE scheme, data are compiled for 114 separate industry branches in Ireland (falling within NACE divisions 1-4), but for reasons of confidentiality the results for some industry branches are aggregated, and at present the census results identify seventy-six 'industries'.

The basic unit in the census is the 'establishment'. A single firm may operate several establishments in different locations, for each of which a separate census return is required; the returns from each establishment are then recorded under the appropriate industrial heading. If a particular firm is highly diversified in its range of production, the returns from individual establishments may be recorded under a number of different census industries. Many individual establishments engage in a variety of economic activities, not all of which are classified within the same census industry. It is not uncommon for individual establishments to manufacture, as subsidiary activities or by-products, commodities which appear as principal products in other census industries. In such cases, an individual establishment can furnish more than one census return, by which means its various activities may be distributed between the appropriate census industries; but in general the establishments included in the census are grouped according to the principal products of each establishment. The summary figures published for each census industry in relation to gross and net output, employment, cost of materials, etc refer to the total value of production of the establishments included in that industry, and may include some activities and products normally attributed to other industries.

Over 3,700 industrial establishments are presently included in the census. Establishments with fewer than three persons engaged are excluded,[5] so not all industrial production and employment is covered. This (necessary) limitation in the coverage of the census is of little consequence in relation to production statistics, such as the estimates of gross and net output in each industry, since it is estimated that the census establishments account for about 90% of all industrial production. It is more important in relation to employment statistics, particularly in certain industries such as apparel, wood and furniture, and metal trades, in which there are a large number of very small firms.

The preliminary reports for each industry summarise the results for all the establishments included in that industry (and record the total number of establishments included). Normally four tables of statistics are published for each industry, and record the gross output of the industry (in value and, where possible, quantity); the cost of materials, fuel, power and heating, packaging materials and similar expenses; the value of net output, and its distribution between salaries, wages and 'remainder of net output', and numbers engaged at a certain date (usually September) broken down by employment, status and sex.

The *gross output* of an establishment represents the net selling value of all goods manufactured in the year, whether sold or not, and the value of work done. This is a straightforward definition, but the actual estimation of gross output may require further elaboration. By definition, it includes (a) the value of goods manufactured and actually sold during the year, (b) goods manufactured and not sold during the year, and (c) work in progress, i.e. partially finished goods at the end of the year.

Consequently, there is generally a difference between the value of sales by an establishment during the course of the year and the value of gross output in that year. Depending on whether there has been a net addition to or net reduction of stocks of finished goods between the beginning and end of the year, the value of gross output will be greater than or less than the value of sales. If sales are taken as the starting point for the calculation of gross output, therefore, and ignoring for the moment work in progress,

gross output =   sales + stocks of finished goods at the end of the year
                 — stocks of finished goods at the beginning of the year

          =   sales + net change in stocks of finished goods

The net change in stocks will be positive or negative (or zero), depending on whether the establishment has added to or reduced its stocks during the year.

Similarly, the net change in work in progress between the beginning and end of the year is added to sales to arrive at gross output. Hence

gross output =   sales + net change in stocks + net change in work in
                 progress.

Work in progress is often very difficult to measure accurately, although in some industries, such as shipbuilding, it may account for a large proportion of gross output over a given period.

There is a further point to be considered. Suppose prices change during the course of the year. If gross output is directly calculated, the change in prices does not matter, for gross output is defined and measured at current values, and we are not concerned here with whether a change in the value of gross output is due to a change in the quantity of goods produced or a change in the prices of these goods.

If, however, gross output is calculated by adding the net change in stocks to the value of sales, as in the formula above, then an adjustment must be made for price changes. The reason is that the value of stocks at the beginning of the year will be different from the actual revenue from the sales of these stocks during the course of the year; if prices are rising, stocks valued at the beginning of the year will realise more than this when later sold. Consequently, if the above formula is used to calculate gross output an adjustment must be made to account for price changes which affect the value of stocks; for example, if prices are rising the value of opening stocks must be re-valued upwards. This adjustment for stock appreciation (or depreciation) is often difficult to estimate accurately.

The *net output* of an establishment is obtained as gross output *less* the cost of materials, fuel, power, etc. bought from other establishments[6] or elsewhere. In many respects net output is a more important figure than gross output. The gross output consists of the value of materials and services bought[7] from other establishments (or imported), plus the 'value added' in processing these materials. A shirtmaking establishment, for instance, buys cloth from textile firms and makes this up into shirts; this 'making-up' process constitutes the value added by the establishment, and forms a proportion of the ultimate value of the finished product. Value added is the difference between the cost of the cloth (and other materials such as buttons, fuel and power, etc.) bought by the establishment and the selling value of the shirts manufactured. Moving one stage further back, the value added by the textile establishment is the difference between the cost of the materials bought and the value of the cloth sold. The final value of the shirt, therefore, is the cumulative total of the value added at each stage of production.

The *value added* by an establishment is distributed in the form of wages and salaries, profits and other payments (such as depreciation and rent) to factors of production which contribute towards the production process. The total of all factor payments in an establishment is therefore equivalent to value added. To the extent that it measures value added, the census net output is a criterion of the importance of an establishment, in terms of its contribution towards the final value of all goods and services produced. In practice the census net output only approximates to value added, since the former includes not only factor payments but certain other payments as well, which do not form part of value added. We discuss this further below.

The other terms used in the preliminary reports, and the statistics compiled, do not require further comment. A specimen report is reproduced in the appendix to this chapter. Certain census industries are treated somewhat differently from the majority, and some notes on these are also included in the appendix.

Industry preliminary reports are published in various issues of the Statistical Bulletin following the census year in question. Following completion of the individual industry reports, principal and summary results for all census industries are published. These include aggregate gross output, cost of materials, net output and persons engaged in each industry; stocks of materials, work in progress and stocks of finished goods held at the end of the year; and changes in fixed capital assets in each industry during the course of the year.

Net output represents approximately the 'value added' by an establishment of which the largest single constituent is wages and salaries. Net output less wages and salaries = 'remainder of net output', including the contribution of factors of production other than labour, i.e. capital, land and enterprise, in the form of profits, interest, rent and depreciation. But net output as defined in the census also includes certain other payments, such as advertising and selling expenses and some other costs which are payments to outside persons or establishments, and hence do not represent 'value added' by the factors of production employed within the establishment. The census does not include any details of these expenses, so that net output only approximates to (i.e. is higher than) the value added by each establishment. In summary, 'remainder of net output' is a residual sum which contains certain supplementary costs not included with costs of materials, but which are not part of value added.

It may be useful to distinguish explicitly between the terms 'gross output' and 'net output' as used in the census of production, and the somewhat similar terms 'gross national product' and 'net national product' (see Chapter 6). If the 'supplementary costs' mentioned above are deducted from the census net output figures, the remainder would be identical with 'value added'. The value added by each establishment, as wages, salaries, profits and other factor payments, represents the contribution of that establishment to the value of gross national product. The total of value added by all census establishments therefore represents the contribution of those establishments to the value of gross national product.

The value added by each establishment includes a provision for depreciation, or capital consumption. If an estimate for depreciation is deducted from value added, the remainder represents the contribution of each establishment to the value of *net* national product. It is important to appreciate the difference between the census definitions of gross and net output and the more recent national accounting terminology.

The results of the census are of considerable interest in analysing changes in the·level and pattern of industrial production and employment over time. Year-to-year changes in the volume of output are normally a compound of changes in the volume of commodities produced and changes in the prices of those commodities. An important aim in analysing the census results is to separate these two components of changes in the value of output, particularly to eliminate the effects of price changes on the value of output, and hence to calculate movements in the *volume* of production from year to year. The time series of gross and net output are not very enlightening; a recorded increase in the value of gross output, for example, may be due solely to a general increase in prices, with no increase in the volume of goods manufactured. The distinction between changes in the value and changes in the volume of production requires the compilation of index numbers, and the methods of construction and use of index numbers to analyse census data are explained below.

The basic unit in the Census of Industrial Production is the establishment. In accordance with EEC regulations, information is now being collected annually for enterprises in addition to that collected for establishments. The enterprise is defined as the 'smallest autonomous unit', which in practice means the company or firm. Within a group of companies each individual company is generally treated as a separate enterprise. Only enterprises engaged wholly or mainly in industrial production and with twenty or more persons engaged in a specified week in September are covered in the annual *Census of Industrial Enterprises.* The return for each enterprise relates to all of its activities and covers all establishments operated by it, including those involved in non-industrial activity. Information is collected on employment, earnings, labour costs, stocks, capital assets, turnover, purchases of materials and fuels, cost of industrial and non-industrial services, taxes, etc. Information is compiled on a NACE basis.

First published results of this census are imminent. There is considerable overlap in coverage and scope with the establishment-based Census of Industrial Production, and to avoid excessive duplication some rationalisation of data-collecting procedures may be desirable.

Index numbers of the volume of production

The object of a volume index of production is to measure the change in the volume of output of a product or group of products over time. The volume of output in a particular period (the 'base' period) is taken as equal to 100, and the volume of output in any subsequent, or earlier, period is related to the base period value of 100. Output in any period is therefore expressed as a percentage of the output in some selected base period. An index is a relative measure, and its value for any period depends upon the base period selected for comparison. Current index numbers of industrial production published by the CSO are related to the base year 1973 = 100, and are calculated on an annual basis; i.e. the volume of output in some recent year — say 1980 — is expressed as a percentage of the volume of output in 1973. The CSO also publishes quarterly and monthly indexes of production. These are considered below.

In practice, the computation of volume index numbers poses a number of practical and conceptual difficulties. The simplest case of a volume index number arises when it is desired to measure the change in output of a single product which is readily measured in physical units. An example might be the output of bricks by a factory which manufactures only one type of standard brick; here the index is based simply on the number of bricks produced. If period 0 is taken as the base period, the index for any other period, say period t, is calculated as

$$\frac{Q^t}{Q^o} \times 100, \text{ where}$$

Q^o = the number of bricks produced in period 0;
Q^t = the number of bricks produced in period t.

Repeating this calculation for each period, a series of volume indexes may be obtained, in which the output in each period is related to the output Q^o in period 0.

In practice, the computation of a volume index is never as simple as this. The volume indexes published by the CSO, for instance, relate to the output of particular industries. Each industry manufactures a wide variety of products, the outputs of which cannot be assumed to change in the same proportion from year to year. To calculate a single volume index for the industry, it is necessary to combine disparate changes in the outputs of different products. How are these individual product changes to be combined? One might take a simple average of the percentage changes in all the products included in the index, but this would be wrong. The shirtmaking industry, for example, manufactures pyjamas as well as shirts, but pyjamas account for at most 10% of the value of the industry's annual output. If the output of pyjamas doubles, whilst the output of shirts remains the same, we cannot take a simple average of these changes and say that the output of the industry has risen by 50%. Nor can we add together the physical outputs of different commodities, for the addition of numbers of shirts and pairs of pyjamas does not result in a meaningful aggregate from which an index can be derived.

We can, however, add together the *value* of the output of shirts and the value of output of pyjamas, giving the total value of output of the industry (i.e. the gross output). This can then be compared with the value of output in the base period. However, a change in the value of output is usually a combination of price changes and quantity changes. To eliminate the effect of price changes on the value of output, the various products of the industry are valued at constant prices. The result is a measure of the value of gross output at constant prices; since prices are held constant, changes in the value of output may be attributed solely to changes in the quantity of products manufactured. A simplified form of volume index may be expressed by

$$V_{ot} = \frac{\Sigma p q_t}{\Sigma p q_o} \times 100$$

Where

V_{ot} is an index of volume of production for period t, with base period $0 = 100$;
q_t are the quantities produced in period t of the various products included in the index;
q_o are the quantities of these same products produced in period 0;
p stands for the set of prices used to value the products included in the index.[8]

The numerator of this index is the total value of the products manufactured in period t, valued at constant prices. The denominator is the value of the products manufactured in period 0, valued at the same constant prices. Since the prices used are the same in both periods, any difference in value must be due to a change in the quantities manufactured. In the formula above, the value of output in period t is expressed as a percentage of the value of output in period 0, and this defines the interpretation of V_{ot}. A volume index for any other period can be calculated by comparing the value of output in that period (valued at constant prices) with the value of output in period 0, valued at the same prices. If the series of index numbers is to be truly consistent, (a) the same set of constant prices must be used to value output in each period and (b) the same period must be used (period 0 in the example above) as a basis for comparison. It is also of course necessary that the same products be included in the index in each period.

In practice, the calculation of volume index numbers presents a number of problems, not least the difficulty — if not impossibility — of obtaining sufficiently accurate statistical data. An industry like metal trades, for instance, manufactures a great variety of products, not all of which can be expressed in homogeneous physical units, and it is virtually impossible to calculate the value of output at constant prices for all the commodities manufactured with the degree of precision suggested by the formula above. There are various ways of dealing with this sort of problem; in general, they involve estimating the volume of production by indirect means.[9] Moreover, as time passes the composition and quality of products manufactured by an industry is likely to change, so that it is no longer possible to maintain an unchanged 'basket of commodities' in the index.

The CSO volume of production index will now be discussed. The basis of the method is the calculation of gross output at constant prices. Suppose it is desired to construct, for a particular industry, a volume index for period 1, with preceding period 0 as base period. Given quantities produced and their prices in both periods, the value of gross output in period 1 can be written as $\Sigma p_1 q_1$ and that in period 0 as $\Sigma p_0 q_0$. The ratio of these two aggregates, viz.

$$G_{01} = \frac{\Sigma p_1 q_1}{\Sigma p_0 q_0} \times 100$$

expresses gross output in period 1 as a percentage of gross output in period 0, and can be described as an index of the *value* of gross output.

To convert this to an index of the *volume* of gross output it is necessary to eliminate the effect of changing prices, which may be done in two ways. The more direct method is to revalue output in both periods using a common set of prices. For instance, period 1 quantities could be re-valued at period 0 prices, giving $\Sigma p_0 q^1$ as the value of output in period 1. The volume index is now derived as

$$V_{o1} = \frac{\Sigma P_0 q_1}{\Sigma P_0 q_0} \times 100$$

Alternatively, period 1 prices could be used as 'weights', with period 0 quantities re-valued at period 1 prices. The volume index would then be

$$V_{o1} = \frac{\Sigma P_1 q_1}{\Sigma P_1 q_0} \times 100$$

The choice of price weights to use in the indexes is not necessarily limited to these two sets of prices, but the two index number formulae written above are the most common. The former, which uses 'base period weights', is known as Laspeyres' Index Number formula, and the latter, which uses 'current period weights', is Paasche's Index Number formula. The weights are prices, and the two indexes differ only to the extent that different sets of prices are used as weights; the quantities are the same in both indexes. This is an important point. Both indexes purport to measure the percentage change in the aggregate value of output at constant prices, in relation to some fixed base period, and changes in value resulting from price changes are eliminated. But except in very unusual circumstances the two indexes will be different; in general

$$\frac{\Sigma P_0 q_t}{\Sigma P_0 q_0} \neq \frac{\Sigma P_t q_t}{\Sigma P_t q_0}$$

even though the quantities used in both indexes are identical. The reason, which need not be elaborated here, is that the relative price weights are different.

The less direct method of estimating a volume index is to 'deflate' the value index by a price index, the result being a volume index. For a single commodity, the proportional change in the value of output from one period to the next is the proportional change in price times the proportional change in quantity. If the volume of output of a commodity increases by 25% and its price increases by 20% the proportional change in value is

$$1.20 \times 1.25 = 1.50$$

Writing G_{ot} as the index of value, P_{ot} as the index of price and V_{ot} as the index of quantity or volume, in general terms the relationship can be expressed as

$$G_{ot} = P_{ot} \times V_{ot}$$

from which it follows that

$$V_{ot} = \frac{G_{ot}}{P_{ot}}$$

In this case the value index (G_{ot}) has been 'deflated' by the price index (P_{ot}) to yield the volume index (V_{ot}).

To calculate a volume index by this method it is first necessary to compute a price index. The methods are analogous to those for volume indexes, with the roles of prices and quantities reversed. For example, a Laspeyres Price Index for period 1, with period 0 as base, is

$$P_{o1} = \frac{\Sigma p_1 q_o}{\Sigma p_o q_o} \times 100$$

while Paasche's Price Index for the same period is:

$$P_{o1} = \frac{\Sigma p_1 q_1}{\Sigma p_o q_1} \times 100$$

In both cases the index measures the change in value of a fixed set of quantities, where Laspeyres' Index uses 'base period quantity weights' and Paasche's 'current period quantity weights'. (For a more extensive discussion of price indexes, see Chapter 8).

Referring back to the earlier formula linking G_{ot}, V_{ot} and P_{ot} with $G_{o1} = \Sigma p_1 q_1 / \Sigma p_o q_o$, and $P_{o1} = \Sigma p_1 q_o / \Sigma p_o q_o$

$$V_{o1} = \frac{G_{o1}}{P_{o1}} = \frac{\Sigma p_1 q_1}{\Sigma p_o q_o} \div \frac{\Sigma p_1 q_o}{\Sigma p_o q_1}$$

$$= \frac{\Sigma p_1 q_1}{\Sigma p_1 q_o} (\times 100)$$

which is a Paasche volume index for period 1. Alternatively, defining $P_{o1} = \Sigma p_1 q_1 / \Sigma p_o q_1$,

$$V_{o1} = \frac{G_{o1}}{P_{o1}} = \frac{\Sigma p_1 q_1}{\Sigma p_o q_o} \div \frac{\Sigma p_1 q_1}{\Sigma p_o q_1}$$

$$= \frac{\Sigma p_o q_1}{\Sigma p_o q_o}$$

which is a Laspeyres volume index for period 1. Note that deflating the value index by a Laspeyres price index yields a Paasche-type volume index, while deflating by a Paasche price index yields a Laspeyres-type volume index.

The price index used by the CSO to deflate the value of gross output is actually the geometric mean of Laspeyres' and Paasche's price indexes,[10] which can be expressed as

$$P_{o1} = \sqrt{\frac{\Sigma P_1 q_0}{\Sigma P_0 q_0} \times \frac{\Sigma P_1 q_1}{\Sigma P_0 q_1}} (\times 100)$$

where the subscript 1 refers to the year for which the index is being constructed (the current year) and the subscript 0 to the previous year. This index is then used to deflate the current year's gross output and hence to derive an index of the volume of production. Written out in full

$$V_{o1} = (G_{o1} \div P_{o1}) \times 100$$

$$= \frac{\Sigma P_1 q_1}{\Sigma P_0 q_0} \div \sqrt{\frac{\Sigma P_1 q_0}{\Sigma P_0 q_0} \times \frac{\Sigma P_1 q_1}{\Sigma P_0 q_1}} \times 100$$

This formula involves a comparison of output in two successive periods. How is the index calculated for subsequent periods, whilst maintaining period 0 as base period? The most obvious method is a direct (binary) comparison between the fixed base period and each subsequent period. By this means the index for period 2 could be calculated as

$$V_{o2} = (G_{o2} \div P_{o2}) \times 100$$

$$= \frac{\Sigma P_2 q_2}{\Sigma P_0 q_0} \div \sqrt{\frac{\Sigma P_2 q_0}{\Sigma P_0 q_0} \times \frac{\Sigma P_2 q_2}{\Sigma P_0 q_2}} \times 100$$

This method can be used to obtain a series of index numbers over time, with a fixed base period = 100.

An alternative method of calculating a series, known as the 'chain-link' method, is to calculate binary indexes for successive periods and then 'chain' the indexes together to form a continuous series. Suppose it is required to calculate indexes for periods 1 and 2, with base period 0 = 100. The index V_{o1} for period 1 is readily calculated, as described above, by means of a direct comparison with the base period 0. To obtain the index for period 2, the first step is to calculate

$$V_{12} = (G_{12} \div P_{12}) \times 100$$

which gives us an index for period 2, with base period $1 = 100$. As it stands, however, this index is not directly comparable with V_{01}, since it is related to a different base period. Th next step is to 'link' the index for period 2 with the fixed base period, by transferring the base from period 1 to period 0. For this purpose it is convenient to regard V_{12} as a ratio rather than convert it to an index by multiplying the above expression by 100. The transfer of the base is then completed by the simple transformation formula

$$V_{02} = V_{01} \times V_{12}$$

and V_{02} is the value of the index for period 2, with base period $0 = 100$.

By the same means the index for period 3 may be found as

$$V_{03} = V_{01} \times V_{12} \times V_{23}$$
$$= V_{02} \times V_{23}$$

where V_{23} is the ratio obtained by means of a binary comparison between the successive periods 2 and 3. Continuing this process of 'chaining' successive indexes together, a continuous series may be derived, with the same base period $0 = 100$.[11]

This second method is the one used by the CSO to obtain series of index numbers for individual industries. The volume index for the current census year is calculated with respect to the preceding year and then linked to a fixed base year (currently 1973). The successive index numbers which arise through the use of this method are not strictly consistent with one another. For instance, the ratio V_{12} is based on a comparison of the quantities produced in periods 1 and 2, whilst V_{23} is based on a comparison of the quantities produced in periods 2 and 3. Although each index is subsequently linked to the fixed base period, there is no direct comparison between period 0 and any subsequent period (except period 1). The successive indexes in the series are not strictly comparable, since they do not involve a common basis of comparison. This does not constitute a serious criticism of the chain-link method, and for all practical purposes the series of index numbers may be regarded as consistent with one another.[12]

Volume index numbers for each census industry are published regularly in the Statistical Bulletin, as part of the analysis of principal results. The indexes are of considerable importance in comparing and analysing the rate of growth of output in different industries.

Although the indexes published by the CSO are referred to as indexes of the volume of production, they are, more correctly, indexes of the value of gross output at constant prices. Although the calculation of volume indexes has been discussed at some length, no rigorous definition of the 'volume of production' has been given. The index numbers which have been discussed, and the ones which are published by the CSO, are index numbers of the volume of gross output. The expression *volume of production* is often taken to mean the volume of gross output, but the two are not necessarily synonymous. An alternative definition of the volume of production is the 'volume of value added' or the 'volume of net output'. If the object in calculating the volume of production is to measure the amount of work done, an index based on net output is the appropriate one to use. Net output is an approximate measure of value added, which represents the contribution of each industry to the value of gross national product.

If the measurement of volume of production is based on net output, then the calculation of an index of volume of production is more difficult. What is required is a measure of the change in the value of net output at constant prices. Suppose an industry manufactures two products only, both of which can be measured in physical units and both of which have a gross output value (price) of £10 per unit. Suppose also that the materials content of product 1 is £8 per unit and that of product 2 is £6 per unit. The net output content or 'net output per unit of output' of the two products are therefore £2 and £4 respectively. In period 0, the industry produces 100 units of each product. In period 1 the industry produces 200 units of product 1 and 150 units of product 2. Using Laspeyres' index number formula, an index of the volume of gross output in period 1 is 175. However if the index is based on net output, quantities in both periods must be weighted by 'net output per unit of output'. Using the same index number formula, this gives

$$V_{01} = \frac{(200 \times 4) + (150 \times 2)}{(100 \times 4) + (100 \times 2)} \times 100$$

$$= 183.3$$

The denominator of this index is the value of net output in period 0. The numerator is the value of net output in period 1, at constant net output prices. The 'net output prices' are the values of net output per unit of output for each product in the base period. The numerator may therefore be regarded as a measure of the volume of net output. There is no reason why the index of the volume of gross output and the index of the volume of net output should be equal in value. They will be equal only if the output of each product changes in the same proportion, or if the relative net output content of each product is the same. The indexes published by the CSO relate to changes in the volume of gross output. This should be remembered in interpreting these indexes.

Also included in the analysis of the results of the census of production are index numbers of the volume of production for certain groups of industries. These comprise:

1   Index numbers for each 2-digit NACE Class (aggregates of NACE 3-digit industry groups).

2   An index for all NACE Classes 11-49 combined.

3   An index for total transportable goods industries (all NACE classes except 16 — Electricity and 17 — Water).

4   An index for total manufacturing industries (transportable goods industries less NACE Classes 11 — Coal Mining, 21 — Extraction and preparation of non-ferrous metal ores, and 23 — Extraction of non-metallic minerals and turf production).

The industry group indexes are calculated as weighted averages of the volume indexes for individual census industries. The aim of the weighting system is to reflect the relative importance of the individual industries comprising each group, and the appropriate weights to use are the respective net outputs of the industries concerned. Details of the method used in Ireland are shown in footnote 13. Following the procedure used for the calculation of volume indexes for individual census industries, group indexes for the current year are first calculated in relation to (i.e. using as base year) the preceding census year. The indexes are then linked to the fixed base year for the whole series (currently 1973).

The monthly industrial production inquiry

The data collected and published in the census are annual and there is a considerable time-lag before the full census results become available. More up-to-date information is provided by a monthly inquiry, the results of which are published some three months after the period to which the inquiry relates. This series began as a quarterly inquiry in 1942, but between 1973 and 1975 was gradually shifted to a monthly basis. The new series of monthly and quarterly production data was first published in the Statistical Bulletin in December 1977 (this issue of the ISB also includes an article describing the basis of the inquiry and the methods of calculation of the production indexes). In 1980 the system of industrial classification was revised in conformity to the NACE system.

The inquiry is based upon a sample of over 1,900 establishments. The size and coverage of the sample ensures that a high proportion of the estimated output of each industry is included in the sample. A principal aim of the inquiry is to collect data on trends in output, and results are published in the ISB under the rubric 'Industrial Inquiries'.

Monthly index numbers of industrial production Monthly index numbers of the volume of production for individual industries and for groups of industries are compiled and published on the basis of the sample returns. Thirty-one individual industry groups are identified.

The index for any industry or industry group in any month is expressed on an 'equivalent annual average' basis — i.e. the monthly production volume is 'grossed up' proportionately to an annual rate, and then compared with the base year volume of production (the present base year is $1973 = 100$). In most cases the volume of production in a particular month is calculated by the following sequence of steps:

(a)   the physical volume or quantity of production in any industry is calculated from the sample returns;

(b)   the physical volume of production is divided by the number of 'standard' working days in the month concerned (the number of 'standard' working days might vary between industries) to yield a daily average volume of production;

(c)   the daily production volume is valued at a given set of prices (usually those for the most recent year for which full Census of Production data are available — this is referred to as the 'link' year);

(d)   the value of production is multiplied by the annual number of standard working days in that industry to give the equivalent annual level of production for the sample respondents;

(e)   the equivalent annual level of production is compared with the annual level of production for the corresponding sample establishments in the 'link' year, and expressed in index number form;

(f)   the index can then be converted to a fixed base year (at present 1973) by means of the 'chain-link' method described above.

A number of points should be noted. First, most industries produce a range of commodities and these have to be individually valued at link-year prices before being added together to give an estimate of industry-wide production. Secondly, the index is based on the production of sample respondents and necessarily excludes non-sampled establishments and those who were sampled but did not respond. Since sample establishments cover more than 90% of persons engaged in Census of Production industries (all establishments with twenty or more persons engaged are included in the sample, and in some industries the limit is fifteen or even ten persons), and the CSO claims a very high response rate (over 97% of establishments sampled in most months), errors in the index due to incomplete coverage are not likely to be serious in most industries and months. The indexes are subject to revision in the light of the later annual census results.

The method described above is not applied universally. Certain products do not easily lend themselves to measurement in standard homogeneous physical units (examples are jewellery, furniture and certain electrical fittings), so less direct methods are used to measure the volume of production. For somewhat different reasons the volume of production of shipbuilding and of printing and publishing are based upon measures of inputs to those industries, rather than outputs. For a more detailed account of these cases and of the general methodology, see the Statistical Bulletin December 1977, pp. 267-8.

Monthly indexes for industry groups are calculated as weighted averages of the indexes for individual branches of industry, the weights being the branch net outputs derived from the census of production results for the most recent year available. Quarterly index numbers of the volume of production for industry groups are also calculated and published. These are calculated as the averages of the corresponding monthly indexes for the relevant three-monthly period.

An important feature of many economic series which refer to specific calendar periods of less than one year is variation due to seasonal factors. Many industries are affected to some degree by seasonal factors such as climate, the availability of supplies at particular periods of the year (e.g. agricultural crops for processing), variations in the pattern of consumer spending, and so on. Comparison of production or production indexes for successive (or different) months or quarters of the year may therefore give a misleading impression of the trend in production in a particular industry or industry group.

Several methods are available by which data may be 'seasonally adjusted' to remove the effects of seasonal factors and thus reveal the underlying trend in a series. The techniques used by the CSO were outlined in Chapter 2. At present a seasonally-adjusted series for the quarterly index of production is published by the CSO for the total Manufacturing Industries index and the Transportable Goods Industries index, but not for individual industry groups or for any of the monthly index series.

Employment, earnings and hours worked While data on production are collected monthly, those on employment and earnings are collected every three months; the sample frame is the same as that for the monthly production inquiry. Information on employment, earnings and hours worked is collected for a particular week in each quarter (mid-March, mid-June, mid-September and mid-December) for each industry branch. These statistics have been collected since 1950. The June 1980 edition of the Irish Statistical Bulletin introduced a new series based on the NACE classification, incorporating a retrospective re-classification back to 1973.[14]

Estimates of employment in a week in each quarter are published for major industrial groups and for total manufacturing and transportable goods industries. These are obtained by grossing-up employment in sample respondent establishments using ratios derived from the most recent available annual census of production data. These estimates are subsequently revised in the light of the census results. For total manufacturing and transportable goods industries, seasonally-adjusted employment series are also published. These vary surprisingly little from the unadjusted series, suggesting that seasonal variations in output may be accommodated more by variation in hours worked than by numbers employed.

A substantial volume of data is published on earnings and hours worked. These series are discussed in Chapter 8 (Prices and Wages), and include average earnings per hour and per week and average hours worked per week in each industry, for all workers and for adult males and females. The series on earnings and hours worked by all workers are also expressed in index number form, the present base of the index number series being September 1973 = 100. Indexes for groups of industries are weighted averages of the indexes for individual industry branches, the weights being aggregate earnings (in the case of the earnings index) and aggregate hours (in the case of the hours index) in the individual industry branches derived from the Census of Production for the base year.

Other series relating to industrial production

Before the introduction of the NACE classification in 1974, the Census of Industrial Production included a number of service-type industries, notably public sector establishments engaged in building and construction activities, and laundries. These establishments are still included in the census population frame, and their results are published in separate tables in the census reports in the Statistical Bulletin. Four groups of establishments are distinguished: local authorities and government departments (building and construction work); canals, docks and harbours (building and construction work); railways (construction, maintenance and repair); and laundries.

To complement the data on public sector building and construction activity there is an annual census of establishments in the private sector, incorporating Building, Civil Engineering, and Allied Trades (NACE Groups 500-504). Results of the *Annual Census of Building and Construction* are published regularly in the Statistical Bulletin. The response rate to this inquiry is poor and the results cannot necessarily be taken as reliable indicators for the building and construction sector as a whole. Published tables include details of work done, materials purchased, payments to sub-contractors, wages and salaries; separate tables with varying degrees of detail are given for small, medium and large establishments. Other tables analyse work done by type of building or activity, and changes in fixed capital assets.

The annual survey of building and construction began in 1966 and was modified in 1975 in accordance with NACE classification. A quarterly inquiry of earnings and hours worked in the building and construction industry, similar to that undertaken for the transportable goods industry, started in 1969. Results of this inquiry are published periodically in the Statistical Bulletin and record average earnings and average hours worked, distinguishing skilled and other (unskilled and semi-skilled) operatives. A monthly index of employment in private building firms employing five or more persons is also published.

Several supplementary sources of data on the output of the building and construction industry exist. The most comprehensive of these is *Construction Industry Statistics* published annually by An Foras Forbartha. Data are culled from a variety of sources, the most important of which are the *Quarterly Bulletin of Housing Statistics,* and the annual *Building Industry in Ireland: Review and Outlook,* both published by the Department of the Environment.

Construction Industry Statistics documents the estimated gross output of the building and construction industry at both current and constant prices. Such data is presented for seventeen sub-sectors of the industry. In each case public expenditure, both capital and current, affecting the industry is identified. In the case of housing, public capital expenditure is disaggregated by spending agency. At the level of the industry as a whole, public and private financing of output are separately distinguished.

The *Annual Report of the Department of the Environment* details the main constituents of Public Capital Expenditure on Housing and documents the number of houses completed during the year. The Quarterly Bulletin of Housing Statistics presents the same data, quarterly, disaggregated by county and county borough, distinguishing public from private housing. The Bulletin also details the current state of the local authority housing programme, the number of dwellings commenced or completed during the quarter, the number under construction, under tender or at the planning stage, and the number of additional dwellings for which sites are currently available, as well as the number of skilled and unskilled employees thus occupied during the quarter.

The *Building and Construction Industry — Review and Outlook* (published annually by the Department of the Environment) presents reviews of seventeen sub-sectors of the industry. Disaggregated estimates of output at both current and constant prices are presented, capital expenditure in each sector of the economy is estimated and the building and construction component identified in each case.

The inquiries described above all relate to past events. Information — inevitably more speculative in nature — on current and short-term future trends in output, employment and so on are collected regularly through the joint Confederation of Irish Industry (CII) — Economic and Social Research Institute (ESRI) *Industrial Survey.* This is carried out quarterly and monthly[15] and covers over 200 establishments in manufacturing industry. Questions are asked on current levels of production, domestic and export sales, employment, stocks and investment; on anticipated levels of production, sales, employment and investment in the following quarter, and on the adequacy of current stocks and productive capacity. Results are analysed by industry sub-group. The survey is a valuable source of information on contemporary trends in industry and on business expectations, and is important as a basis for short-term forecasting.

Appendix

MANUFACTURE OF PAINTS, VARNISHES AND PRINTING INKS

(N.A.C.E. Group No. 255)

Number of establishments: 21 in 1975 and 22 in 1976.

I. — GROSS OUTPUT

Goods manufactured		Quantity		Net selling value	
		1975*	1976	1975*	1976
				£	
Enamels	litres	4,251,740	4,749,498	3,280,766	4,209,522
Varnish, oil or spirit (incl. stains, lacquers and liquid dryers)	,,	1,126,567	1,371,059	670,282	806,854
Oil paints, mixed, ready for use (incl. under coats)	,,	2,820,467	3,001,768	2,007,294	2,328,769
Emulsion paints	,,	4,269,351	5,108,016	2,987,514	3,937,460
Other paint and varnish products (incl. distemper)		(value only)		831,699	902,979
Inks		,,		1,557,633	2,151,568
Other products and work done		,,		1,947,168	4,525,401
Work in progress at end of year other than in above headings		,,		183,085	84,953
Total value of goods made and work done				13,465,441	18,947,506

II. — MATERIALS USED

Description		Quantity		Cost	
		1975*	1976	1975*	1976
				£	
Resins	tonne	7,805	6,972	2,339,300	2,162,327
Pigments or dry colours	,,	6,696	6,151	2,007,640	1,261,010
White spirit	litres	2,441,623	2,178,107	215,668	241,651
Varnish	,,	124,640	150,807	63,470	82,820
Solvents		(cost only)		497,004	805,849
Oils, refined and unrefined		,,		752,469	1,049,763
All other materials		,,		980,978	3,941,641
Work in progress at beginning of year other than in above headings				19,179	50,920
Total cost of materials				6,875,708	9,595,981

III. — NET OUTPUT, SALARIES AND WAGES

		1975*	1976
		£	
(1)	Gross Output	13,465,441	18,947,506
(2)	A. Cost of materials	6,875,708	9,595,981
	B. Cost of fuel, etc.	200,546	273,994
	C. Cost of packaging, materials for repairs, etc.	1,418,242	1,985,607
	Total cost of materials, etc.	8,494,496	11,855,582
Net Output = (1) — (2) = (3) + (4)		4,970,945	7,091,924
	Salaries	1,083,055	1,288,066
	Wages and earnings	1,269,200	1,539,958
(3)	Total salaries and wages	2,352,255	2,828,024
(4)	Remainder of Net Output	2,618,690	4,263,900

IV. — NUMBER OF PERSONS ENGAGED IN A WEEK IN SEPTEMBER, 1975 AND 1976

Category	Number of persons			
	1975		1976	
	Males	Females	Males	Females
Industrial workers (excluding apprentices and basic supervisory staff)	461	74	464	53
Apprentices	1	—	1	—
Basic supervisory staff (including foremen, supervisors, etc.)	35	2	38	2
Total	497	76	503	55
Administrative, clerical and technical staff	247	133	251	139
Total number engaged	744	209	754	194

*Revised.

Source: Irish Statistical Bulletin, June 1980.

(a)   For some industries Table I and/or Table II are curtailed to give only the total value of gross output and/or the total cost of materials used, to prevent the identification of individual firms.

(b)   For some industries (e.g. sugar, brewing) the statistics do not relate to the calendar year but to the nearest over-lapping twelve-month period.

(c)   The reports for some industries (e.g. electricity generation, gas) include additional data on the value of construction, repair and maintenance work carried out by establishments in the industry.

Notes to Chapter 4

[1]   *Final Report of the first Census of Produciton of the United Kingdom* (1907) Ed. 6320, 1912 (HMSO).
[2]   The figures related to the whole of Ireland and the census was very limited in scope.
[3]   *P. No. 844* (Stationery Office, Dublin). Preface, p. iii and p. xxi.
[4]   For comparison of the two systems, and an explanation of the differences, see the *Irish Statistical Bulletin* September 1979.
[5]   For two industries (bread, wood and furniture), in which small establishments are numerous, only simplified returns are required for establishments with three to nine persons engaged; as a result some of the statistics relating to these industries refer only to concerns with ten or more persons engaged.
[6]   Both gross and net output are defined and measured in relation to the basic census unit, the establishment. The gross and net output for individual industries or groups of industries, as published in the preliminary reports and other tables, are obtained simply as the aggregates of gross and net output for the individual establishments included in each industry.
[7]   The figures published for 'costs of materials' etc. actually refer to the value of materials used by the establishment during the course of the year. This may be different from the value of materials purchased by the establishment in the course of the year, since firms may add to or reduce stocks of materials.
[8]   Written out in full, the numerator of the index would be

$$p^1q^1_t + p^2q^2_t + p^3q^3_t + \ldots p^nq^n_t$$

and the denominator would be

$$p^1q^1_0 + p^2q^2_0 + p^3q^3_0 + \ldots p^nq^n_0,$$

where the superscripts denote the products (1,2,...n) included in the index.
[9]   The index could be based upon the change in output of a limited number of the most important products, on the assumption that the output of all other products changes in roughly the same proportion. Alternatively, an estimate of the total value of output at constant prices could be obtained by 'deflating' the current value of output by a general price index. A combination of both these methods may be used. A rather different approach is to assume that the volume of output changes in direct proportion to some other measureable factor. For example, if a constant conversion rate could be assumed between the input of turf (in tons) and the output of electricity, the volume of output of turf-fired electricity-generating stations could be estimated on the basis of the quantity of turf used. Similar methods might be used in other cases, where it is reasonable to assume a simple and direct relationship between the input of a particular material and the output of the industry. However, a simple input structure usually implies a simple output structure, in which case the volume of output can be measured directly. If it is difficult to measure the volume of output directly, it is usually also difficult to measure the volume of input. An analogous procedure is to base a measure of changes in the volume of output

upon a measure of changes in labour input. It is not very easy in most cases to measure changes in labour input, since there are factors such as changes in man-hours worked and changes in productivity to be accounted for. In general, measures of production based on inputs — whether of materials or of labour — are much more useful in service-type industries in which outputs cannot, by their nature, be measured in physical units.

[10] This is known as Fisher's 'Ideal' index.

[11] As a numerical example, suppose output rises by 50% between period 0 and 1, 20% between period 1 and 2, and 10% between period 2 and 3. Then $V_{01} = 150$; $V_{02} = 150 \times 1.20 = 180$; $V_{03} = 180 \times 1.10 = 198$.

[12] This objection does not apply to the first method described, since the base period quantities used for successive indexes remain fixed. It is worth noting that V_{ot} as calculated by this method is not in general equal to V_{ot} as calculated by the chain-link method.

[13] For a comparison between any two periods, say periods 0 and 1, there is a choice of using either period's net outputs as weights. As with the price index referred to earlier, the CSO uses a Fisher-type index number formula, involving the net outputs of both periods. This can be expressed as:

$$V^{*}_{01} = \sqrt{\frac{\sum_{i=1}^{m} V^{i}_{01} q^{i}_{o}}{\sum_{i=1}^{m} q^{i}_{o}} \times \frac{\sum_{i=1}^{m} V^{i}_{01} q^{i}_{1}}{\sum_{i=1}^{m} q^{i}_{1}}} \quad (\times 100)$$

Where

V^{*}_{01} is the combined index for m industries

V_{01} is the volume relative (i.e. volume index divided by 100) for industry i ($i = 1, 2 \dots m$)

q^{i}_{o} is the net output of industry i in period 0

q^{i}_{1} is the net output of industry i in period 1

[14] The retrospective re-classification to NACE was carried out for each September quarter for 1973-76, and for each quarter of the year from March 1976.

[15] The monthly survey was inaugurated in 1974, when the survey was revised to obtain greater comparability with similar surveys in other EEC countries. See P. Neary, 'The CII-ESRI quarterly and monthly surveys of business attitudes: methods and uses' *ESRI Quarterly Economic Commentary* March 1975.

5 Foreign Trade

For Ireland, trade with other countries is extremely important, since a large proportion of her raw materials and capital goods, as well as finished goods for current consumption, are imported. In order to pay for these imports it is necessary to export a proportion of domestic output, and the balance of payments which records, inter alia, the relation between total imports and total exports of goods and services, is an important factor in economic policy. Imports now account for around 60% of expenditure on gross national product,[1] and this implies that roughly the same proportion of domestic output must be exported, or that alternative means of financing these imports must be found.

We begin by distinguishing *merchandise trade* from *non-merchandise trade*. The former relates to trade in physical commodities and the latter covers trade in services, such as tourist expenditure, earnings from insurance, banking, shipping and miscellaneous other services. Merchandise trade accounts for the major part of total trade (both imports and exports), but non-merchandise trade is often very important, particularly in Ireland and also in the United Kingdom, in that it may compensate for deficits in the merchandise trade balance. For example, in 1978 Ireland's merchandise exports of £2,917.0m fell well short of merchandise imports, which amounted to £3,646.9m. This deficit, however, was greatly reduced by the surplus on non-merchandise items — exports (credits) were £1,254.7m and imports (debits) £673.6m. Non-merchandise items, e.g. receipts from tourism and income from investment abroad, are important sources of foreign exchange earnings for Ireland.

Changes in the value of trade are a compound of changes in volume and changes in prices. It is possible for a country's trading position to worsen solely through relative price changes in imports and exports, even though the volume of trade has not changed. If, for example, the prices of imported goods and services rise whilst the prices of exports remain stable, a greater volume of exports is required to pay for the same volume of imports. Changes in the relative prices of imports and exports and measures of changes in the 'terms of trade' are discussed in Chapter 8.

The annual value of all merchandise and non-merchandise imports constitutes total imports on current account, and similarly for exports. If total current imports exceed total current exports, there is a *deficit* in the balance of payments current account. An excess of total exports over current imports constitutes a *surplus*. The size of this deficit or surplus, and its variation from year to year, is an important factor in the formulation of domestic economic policy. Generally a surplus is considered favourable and a deficit unfavourable, though it does not follow that a country should strive each year to achieve a surplus in its balance of payments.

The balance of payments also includes a statement of capital transactions between Ireland and the rest of the world, such as loans raised abroad by the Irish government, investment by foreign residents, and similar capital transfers. It will be explained later how a deficit or surplus in the balance of payments current account is always matched by an equivalent surplus or deficit on capital account. Taking current and capital items together, the balance of payments accounts always 'balance', i.e. total credits = total debits, in the way in which a company's trading accounts always balance. When reference is made to a deficit or surplus in the balance of payments, this normally refers to current account only, though it may also be used with reference to capital account only. In an accounting sense, however, the sum of all current and capital debit items (imports) must equal the sum of all current and capital credit items (exports). In the case of Ireland, there is normally a deficit on current account (and hence a surplus on capital account), and it is the magnitude of this deficit which is important in economic policy.

Merchandise trade

Statistics of merchandise trade are available in considerable detail, mainly in the monthly *Trade Statistics of Ireland,* compiled by the Central Statistics Office from data supplied by the Revenue Commissioners. Statistics of merchandise trade have been systematically collected and published since 1924. Since then, there have been various legislative changes affecting the collection and presentation of the statistics, and it is generally not possible to obtain consistent series for individual components of trade over the whole of this period. Nevertheless, as a result of the operation of Customs and Excise regulations, statistics of merchandise trade are extremely comprehensive, and consistent series for many aggregates are available from 1924. The present discussion of foreign trade is based upon the currently published statistics, with historical references where necessary.[2]

The statistics of merchandise exports and imports are derived from declarations made to Customs officials, required by law for all exports and imports. The published statistics are affected by certain legal and administrative regulations, which should be remembered in interpreting the figures.[3] For instance, certain commodities are excluded from the published trade statistics:

(a)   personal and household effects and parcels brought by passengers for private use;
(b)   fish arriving direct from territorial waters;
(c)   live animals temporarily imported or exported for racing, breeding or show purposes only;
(d)   ships' stores and bunkers for use on the carrying vessel and ballast of no commercial value;
(e)   goods imported by the representatives of foreign governments and goods exported to Irish government representatives;
(f)   imports of a charitable nature from organisations such as the Red Cross Society, and exports, by way of relief, of goods purchased in this country by the government or by Irish charitable organisations;
(g)   works of art temporarily imported or exported for exhibition only;
(h)   transit trade;
(i)   currency and notes and silver or base metal coin, being legal tender.

The basis of valuation of imports and exports is also important. Merchandise imports are valued c.i.f. (cost, insurance, freight) at point of entry, i.e. the stated value of the goods, including insurance and freight costs incurred in shipment of the goods to the point of entry (but excluding customs duties). Exports, on the other hand, are valued f.o.b. (free on board) at the point of exit, i.e. the net selling value of the goods at point of exit, exclusive of insurance and freight charges for further shipment abroad (and exclusive of any subsidies or drawback). The published figures of merchandise exports and imports cannot therefore be compared on a strictly consistent basis, since the latter includes a certain element of 'non-merchandise' trade. Moreover, some of these insurance and freight costs may be paid to Irish shippers and insurance agents, thus forming part of Ireland's invisible export earnings.

The points of detail and qualifications discussed above are particularly relevant when comparisons are made of monthly or annual levels of merchandise imports and exports. The difference between merchandise imports and merchandise exports is described as the 'balance of trade' or the 'trade gap'. If imports exceed exports there is an import excess, whilst the reverse situation constitutes an export excess; sometimes the expressions 'adverse' and 'favourable' balance of trade are used (though not by the CSO) to describe an import or export excess, but this terminology can be misleading and should be avoided.[4] Moreover, the balance of trade relates to merchandise trade only and trends in the balance of trade cannot be taken to mean that similar trends are occurring in the overall balance of payments.

Another factor to be taken into account in analysing the trends in monthly merchandise trade is seasonality. The effects of seasonal trends on the value of imports and exports are discussed below.

Trade by commodities and groups of commodities

Two broad classifications are used in the presentation of monthly trade statistics: trade by commodities and trade by countries.

The basis of the commodity classification used in the tables of Trade Statistics is the United Nations Standard International Trade Classification (SITC). The SITC includes ten major sections,[5] each section being further sub-divided into divisions. Sections are designated by a one-digit code (0-9) while divisions are designated by a two-digit code. A section may contain up to ten divisions.[6] Each division is sub-divided into groups and sub-groups of commodities at three, four and five-digit level, the last representing the typical minimum level of commodity classification in the SITC.[7] The use of the SITC facilitates international comparisons of trade between countries using it, at a fairly detailed level. Its introduction involved some extensive changes in the system of commodity classification used in Ireland before 1963, and it is difficult to derive consistent series at the commodity level for the overlapping period.

The introductory tables of the monthly trade statistics summarise the value of trade in total, and by sections and divisions of commodities.

The values of imports and exports are recorded for the current month and for the corresponding month of the previous year. Also recorded are the cumulative totals from January of the current year, and the corresponding cumulative totals for the previous year. Thus, for June 1980 the figures include the value of trade in that month, the value of trade in June 1979, and the cumulative value of trade for January-June 1980 and January-June 1979. The December issue includes the cumulative totals for the twelve months of the year, and these stand as provisional estimates of trade for the year.

Later tables in the monthly publication record merchandise imports and exports in much greater commodity detail (e.g. at 4-, 5- or sometimes 6-digit level). The statistics in these tables are presented on a similar basis to those in the summary tables, recording trade in the current month as well as the cumulative calendar year totals for the current year and the previous year. In addition these tables record, where relevant, details of the quantity as well as the value of imports and exports. The figures record total imports and exports for each 5-digit commodity group; the value and volume of trade with individual countries is given where the value of the trade flow exceeds £20,000 for the month in question, or exceeds an average of £15,000 per month for the year in question. Subject to the same proviso, trade with the EEC member countries is also recorded.

Imports to and exports from Shannon Free Airport, which is a customs-free zone, are treated separately from other trade. Firms operating in Shannon are exempt from normal customs requirements and documentation. The values of imports to and exports from Shannon Free Airport, which are available from special returns submitted directly to the CSO, are not included under their respective commodity headings in the detailed tables of imports and exports at 5-digit level. Instead, the total value of imports to and exports from Shannon are shown under a separate heading in SITC Section 9, and no commodity detail is available. In the analysis of trade by country (see below), the value of Shannon trade is included only for trade with EEC countries, the USA and Canada.

Trade by countries
The breakdown of trade by country in the monthly accounts follows a similar pattern. A summary table records the value of total trade with each country (nearly 200 are listed) for the current month, and the cumulative calendar year total.

It is important to understand the system used for classifying imports and exports under different country headings. In this country imports are classified according to 'country of origin', which is that country in which the imported products were grown, processed or manufactured. Exports are classified according to 'country of final destination'. The country of origin or final destination may be different from the country of consignment or shipment. All three may be different; an export order may be consigned to Hong Kong, shipped to Liverpool, and ultimately consumed in China. The system of classification by origin and final destination is to be preferred, though it may not be easy or possible in every case to distinguish between, say, consignment and final destination.

Following the summary table of total trade by country, a subsequent table records the value of trade with each country at 2-digit (division) level. This represents the most detailed published analysis of trade by countries.[8] Other summary tables show the value of trade by area, including Great Britain, Northern Ireland, other EEC countries, EFTA countries, Eastern European countries, North America, and some other groupings.

These tables which record the commodity pattern of Ireland's trade and trade relations with different countries are of great value in analyses of the structure of Ireland's international trade. The monthly publications also include certain aggregate tables which are of interest in analysing general trends in trade. These include a summary table which records the value of imports and exports and the balance of trade for each month for the preceding 3-4 years. Other important tables record monthly and annual index numbers of the volume of trade, import and export price index numbers and seasonally adjusted estimates of the value and volume of trade.

Two other summary tables published in the monthly Trade Statistics are of interest. The first presents a breakdown of imports by value, for the current month, the cumulative total for the current year, and the cumulative total for the corresponding months of the previous year, and by use, distinguishing producer's capital goods, consumption goods, and materials for further production (intermediate goods). Such a breakdown is useful for analytical and policy purposes — for example, a rise in the trade deficit caused by a rapid rate of growth in imports of capital goods or intermediate goods may imply different policy responses from a deficit caused by a rise in imports of consumption goods. Moreover, in constructing models of the economy it is frequently desirable to estimate separate equations for each category of imports.

The second table analyses exports according to broad sector of origin, identifying agricultural produce, forestry and fishing produce (this category is very small) and industrial produce. This goes some way towards distinguishing between natural resource-based and non-resource-based exports, and allows for the possibility of fitting separate equations for these categories of exports in models of the economy.

Index numbers of the volume of imports and exports

Changes in the value of trade are a compound of changes in the volume of trade and changes in prices. In analysing the trends in merchandise trade it is useful to identify the separate influence of these two factors on the value of trade. For instance, to what extent is an increase in the value of merchandise imports due to a rise in import prices, and to what extent to an increase in the volume or quantity of imports?

The main tables in the monthly publication record the quantity as well as the value of imports and exports. For individual commodities it is easy to devise index numbers which measure the change in the volume of imports or exports. More generally, the aim is to construct an index or number of indexes which will measure the changes in the volume of all commodities or of groups of commodities taken together. The problem here is similar to that discussed in the previous chapter, where it was required to construct an index to measure the change in the volume of output of a number of products taken together. This was achieved by measuring the change in the value of output at constant prices. Similar methods are used to construct indexes of the volume of merchandise imports and exports. Imports and exports are revalued at constant prices, and the resulting series of constant price aggregates converted to index number form. Volume index numbers of imports and exports are published in the monthly trade publication and in the Irish Statistical Bulletin.

Monthly volume index numbers for total imports and exports are published in each issue of Trade Statistics, which records not only the indexes for the current month but also, for comparative purposes, the monthly indexes for the preceding three years. The base period for these indexes is (at present) 1975, i.e. the volume of trade in the current month is related to the average monthly level of trade in 1975.

These monthly indexes may be used to analyse the trend in the volume of merchandise exports and imports from month to month. It is important to allow for seasonal variation in the pattern of trade; like many other economic variables, changes in the monthly pattern of imports and exports are affected by seasonal factors which are independent of the underlying trends in trade.[9] These seasonal influences must be eliminated in order to determine the underlying trends in imports and exports.

Seasonally-adjusted estimates of the value of trade, and seasonally-adjusted volume indexes of imports and exports, are published in the monthly Trade Statistics. The adjusted series 'smooth out' the effects of seasonal factors on the monthly figures and reveal more clearly the underlying trends in the series.[10] The seasonally-adjusted series are, however, still subject to non-systematic irregular and cyclical influences and should not be regarded as unambiguous measures of trend, especially over relatively short periods of time.

Annual indexes of the volume of imports and exports are also recorded in the monthly Trade Statistics and in the Statistical Bulletin. These are not the average or arithmetic mean of the monthly indexes, but are independently computed.

The general aims and methods of construction of volume indexes were discussed in some detail in the previous chapter, and here only a brief description will be given of the method of construction of the import and export volume indexes. It is sufficient to understand the logic of the techniques used.

The most direct method of constructing a volume index for a group of commodities is by calculating the aggregate value of the commodities in each period at constant prices, and then converting these constant value aggregates into index number form. However, an alternative method of re-valuing imports and exports at constant prices is by deflating the current values of imports and exports by suitable import and export price index numbers. For example, if the volume of imports rises by 20% between periods 0 and 1, and the prices of imports rise by 50%, the value of imports will rise by 80%. This ratio may be obtained arithmetically as $1.20 \times 1.50 = 1.80$. Consequently, given the change in the value of imports, and given the change in the import price level, the volume of imports can be estimated by dividing the current value of imports by the import price index (i.e. $1.80 \div 1.50 = 1.20$). By this means the current value of imports is 'deflated' to obtain the value of imports at constant prices. The 'constant prices' will depend upon the base year for the price index number; if, say, the base year for the import price index is 1975 = 100, then the constant value aggregates will be expressed at 1975 prices. To express this symbolically, and in a simplified form, let $\Sigma \, p_t \, q_t$ represent the value of imports in period t, at current period t prices, $\Sigma \, p_0 \, q_t$ the volume of imports in period t (i.e. the value of imports at base period 0 prices) and P_{ot} the import price index for period t, with base period 0 = 100. Then the volume of imports can be estimated as

$$\Sigma \, P_0 \, q_t \; = \; \frac{\Sigma P_t \, q_t}{P_{ot}} \; \times \; 100$$

The result is an estimate of the value of imports in period t at constant (period 0) prices. Repeating this calculation for each period gives rise to a series of estimates for the value of imports at constant prices, which may then be converted to index number form. Volume indexes for exports may be similarly calculated by deflating the value of exports at current prices by corresponding export price index numbers.[11]

Annual and monthly volume indexes of imports and exports in Ireland are calculated by similar means. Current value figures for imports and exports are 'deflated' by import and export price indexes to obtain estimates of the value of trade at constant (currently 1975) prices. These constant value aggregates are then converted into index number form by relating the value of imports/exports in each month to the average monthly value of imports/exports in the selected base year. For the annual volume index, the constant price estimate of imports/ exports in the current year is related to the level of imports/exports in the base year. There are certain disadvantages in calculating volume indexes in this way, but the method is frequently used and the results may be regarded as quite reliable.[12]

One point of interest here is the relation between the monthly and the annual volume indexes. These are independently calculated, and the annual index may differ — sometimes substantially — from the average of the twelve monthly indexes. The reason is that the price indexes used to deflate the monthly current values of imports and exports are independent of those used to deflate the annual figures. The annual price index employs a different regimen and weighting system from that used to calculate the monthly index, and the index number formulae used are different (see Chapter 8).

The volume indexes referred to above relate to total exports and total imports only. No volume indexes are published for specific groups of commodities or for trade with specific countries or groups of countries.

The monthly and annual volume index series for imports and exports are also published in the Economic Indicator Series in each issue of the Statistical Bulletin. The monthly trade statistics and the Economic Indicator Series also include the import and export price indexes which are used to derive the volume index series. The methods of calculation of these price indexes are discussed in Chapter 8, which also explains how the import and export price indexes are used to calculate changes in the 'terms of trade'.

The balance of payments

The balance of payments is a summary statement of all transactions between one country — in this case Ireland — and the rest of the world, over a certain period of time. For countries in which exports and imports constitute a large share of output and expenditure, the annual balance of payments statement is of considerable interest, and trends in the balance of payments exert an important influence on economic policy. Estimates of the balance of payments are published in the Irish Statistical Bulletin, and in less detailed form in the annual *National Income and Expenditure Accounts.* A modified version of the 1978 balance of payments statement is shown in Table 5.1.

A distinction can be made in the balance of payments statement between 'current' and 'capital' items. The former include all receipts and payments for goods and services bought from or sold abroad in the course of the year, as well as certain international transfer payments (pensions, emigrants' remittances) and factor incomes received from or paid abroad. Capital items include all transfers of capital funds or asset-holdings, in the form of loans, loan repayments, purchase and sale of securities and similar transactions, between this country and the rest of the world. This distinction may be further clarified by examining, item by item, the transactions recorded in the balance of payments statement shown in Table 5.1.

Current items
Merchandise trade. This, the biggest single item in the accounts, has been discussed in some detail already. The difference between exports and imports is the 'balance of trade', and for Ireland there is invariably a deficit in the merchandise trade balance — debit items (imports) exceed credit items (exports). There is a difference between the figures for

merchandise exports and imports recorded in the balance of payments accounts and the corresponding figures published in the Trade Statistics. The former are 'adjusted for balance of payments purposes'. These adjustments include the deduction of some temporary transactions and re-directed trade from both exports and imports (since the goods concerned do not change ownership and no process of transformation takes place), and the deduction from exports of goods purchased and exported for storage abroad under EEC intervention schemes. Actual sales from these foreign stores are, however, added to exports. Other small adjustments made include revisions to earlier-published trade statistics.

Table 5.1 Balance of International Payments, 1978

	£m Irish	
	Credits	Debits
A. Current flows		
1. Merchandise trade	2,917.0	3,646.9
2. Freight and other transportation	175.7	82.2
3. Tourism and travel	215.9	178.5
4. Other services	80.9	21.8
5. Income from capital	230.2	310.3
6. International transfers:		
(a) Emigrants' remittances	49.4	0.8
(b) Pensions and allowances	47.9	4.6
(c) European Communities	410.2	67.4
(d) Other private and government transfers	19.2	8.0
7. Balance unaccounted for	25.3	—
Total current items	4,171.7	4,320.5
Net balance on current account (deficit)	148.0	
B. Capital flows		
8. Private capital	198.8	361.3
9. Government capital	338.2	—
10. Central Bank transactions (net)	—	54.6
11. Other banking, etc. transactions (net)	12.4	
12. Capital transfers	15.3	—
Net balance on capital account (surplus)		148.8

Source: *Irish Statistical Bulletin,* December 1979.

Non-merchandise current items. The remaining items on current account represent receipts and payments with respect to services, factor incomes and international transfer payments. Such statistics are more difficult to compile than those for merchandise trade, and are published less frequently and in less detail.

Item 2 includes the total receipts (foreign and domestic) of Irish sea and air carriers arising from the import of goods into Ireland, and total foreign receipts arising from all other international shipments of freight and carriage of passengers. Merchandise imports are valued c.i.f., and this includes payments by importers to Irish freight carriers. The resulting overestimate of foreign payments is offset by including these payments as exports of freight services on the credit side of the accounts.

Item 3 records receipts and expenditures with respect to tourism and travel. On the credit side this represents expenditure by visitors (tourists, businessmen and other visitors) in Ireland, which entails expenditure on Irish goods and services by foreign residents and is thus a form of export. On the debit side, expenditure by Irish residents travelling abroad is a form of import. Expenditure by visitors to this country is generally higher than expenditure by Irish travellers abroad, and has been a major source of export earnings for Ireland, but its relative importance has declined substantially in the last decade.

Estimates of expenditure by visitors to Ireland and by Irish residents travelling abroad are published annually in the Statistical Bulletin. The estimates are based on a sample of travellers, both incoming and outgoing, covered by a Passenger Card Inquiry. Information is collected on reason for journey, length of stay, expenditure during stay and normal country of residence, and this information is used in estimating total expenditure by tourists and other travellers. These estimates are subject to the usual sampling error, but it is reasonable to assume that they are satisfactorily accurate for balance of payments purposes. For further discussion of the Passenger Card Inquiry and its limitations, see Chapter 7.

Item 4 (other services) contains a miscellany of items including expenditure by foreign diplomatic personnel in Ireland (a credit) and expenditure by Irish diplomats abroad (a debit), commission earnings of import agents (a credit), rental payments on imported films (a debit) and a variety of other services.

The largest non-merchandise item on current account is 'income from capital' (item 5). On the credit side, this includes the receipt by Irish residents, companies and other institutions (including government and banks) of profits, dividends and interest on assets held abroad.[13] On the debit side, there is a similar outflow of payments made by Irish-based companies and public authorities to foreign residents, companies and governments owning assets in Ireland. Estimates for the value of these receipts and payments are based upon data provided by the revenue commissioners, the Central Bank, the commercial banks, insurance companies and similar financial institutions.

Items 6 (a) and 6 (b) require little elaboration. A high rate of emigration from Ireland has resulted in a large inflow of remittances from emigrants to families resident in Ireland, and these transfer payments from abroad are included as part of export earnings. Their importance has declined substantially in recent years. Item 6 (b) consists principally of pensions and allowances paid by the British Government to former civil servants and members of the armed forces resident in Ireland. Debits under both these headings are very small. Receipts and payments under these headings are estimated from a variety of sources and are subject to a statistically imprecise margin of error. These sources include British and other foreign postal and money orders cashed and British and American banknotes taken in by the commercial banks, which mainly represent emigrants' remittances.

Item 6 (c) incorporates the value of payments received from and paid to various EEC Funds, including on the credit side receipts from the Agricultural Guidance and Guarantee Fund, the Social Fund, and Monetary Compensation Amounts which arise from one of the more tortuous aspects of the Common Agricultural Policy. Debits include the Irish contribution to the EEC budget.

Item 6 (d) includes a variety of private and public transfer payments, e.g. the estimated value of gifts received by Irish residents from friends or relations abroad (a credit), and the value of gifts sent abroad (a debit).

The sum of items 2-6 constitutes trade in services (items 2-5) and current transfers (item 6), and for Ireland there is normally a surplus balance on these components of non-merchandise trade — that is, export or credit items always exceed import or debit items. This surplus helps to offset the deficit on merchandise trade (item 1).

The remaining item on current account is the 'balance unaccounted for'. This is a residual, the magnitude of which is fixed to ensure that the deficit (or surplus) on current account is exactly equal to the surplus (or deficit) on capital account. It is a necessary condition of the balance of payments that the sum of all credit items (current and capital) should equal the sum of all debit items. If the sum of all credit items on capital account is found to exceed the sum of all debit items on capital account by, say, £x million, it necessarily follows that the sum of all debit items on current account must exceed the sum of all credit items by the same amount. Hence a surplus (or deficit) on capital account must be offset by an equal and opposite deficit (or surplus) on current account. In Ireland, it is the practice to treat any 'balance unaccounted for' arising from the construction of the accounts as current in nature. In short, the item 'Balance unaccounted for' is included as part of current receipts and payments to ensure that the deficit on current account equals the surplus on capital account. This does not necessarily mean that estimates of capital transactions are statistically more reliable than estimates of current transactions; in fact the balance unaccounted for incorporates errors and omissions in the estimates of transactions on current and capital account.

Adding together the value of all current exports of goods and services and inward current transfers, for 1978 this total (£4,171.7 m) fell short of the corresponding total (£4,320.5 m) of current imports of goods and services and outward current transfers by £148.8 m. This deficit in the current balance of payments for 1978 is entered as a credit item in the balance of payments accounts, thus ensuring that the sums of credit and debit items are equal. A surplus in the current balance would be entered as a debit item.

Capital items

The capital account is designed to record the flow of capital funds into and out of Ireland. A flow of funds into Ireland is treated as an export or credit item, and a flow or transfer of capital funds out of Ireland is treated as an import or debit item. There is sometimes difficulty in understanding whether a particular transaction should be treated as a debit or credit item, particularly in view of the widespread use of the terms 'export of capital' and 'import of capital' — these are respectively debit and credit items in the balance of payments. The distinction between debit and credit items will be clarified by examining the capital transactions recorded in Table 5.1.

Item 8, private capital flows, includes a wide variety of non-governmental transactions, of which the major component on the credit side is portfolio and direct foreign investment in Ireland in the course of the year; correspondingly, portfolio and direct investment abroad by Irish residents (individuals or companies) represents a debit under this heading. Other transactions include premiums on life assurance paid by Irish residents to foreign-established insurance companies; changes in the net deposits of non-residents at Irish banks, and changes in the net deposits of residents at external branches of Irish banks. In 1978 there was a substantial net outflow of private capital; the balance of private capital transactions, however, tends to vary markedly from year to year.

Item 9 records government borrowing from abroad (excluding any borrowing from the International Monetary Fund, which appears in item 10) through the issue of short-term or long-term stocks, and is related to the balance between government revenue and expenditure and the public sector borrowing requirement. Borrowing from abroad is a credit item in the balance of payments, while repayments of earlier loans are debit items; the figure for net borrowing in any year represents the net change in the government's foreign indebtedness. The interest paid on foreign borrowing will appear as a current item in the balance of payments under the heading 'income from capital' (item 5 in Table 5.1).

Item 10 records the *net* change in external assets held by the Central Bank, and the figure of £54.6m in Table 5.1 implies that these external assets *increased* by this amount during 1978. An increase in the Central Bank's external assets can be thought of as lending abroad or net capital outflow, which is a debit item, while a fall in the Bank's external assets can be thought of as borrowing from abroad or net capital inflow, which is a credit item in the balance of payments.

The external assets of the Central Bank constitute the official external reserves of the country and are held in various forms: foreign currencies and securities (dollars, sterling, etc.), gold, subscriptions and lending to the International Monetary Fund (IMF), and Ireland's allocation of Special Drawing Rights (SDRs), which are international reserve assets created by the IMF. External liabilities of the Central Bank include Irish currency held by the IMF, external banks and other non-resident institutions.

Item 11 records the net change in the external position of commercial banks and some other financial institutions (such as hire purchase companies) incorporated in Ireland. This includes the external lending activities of banks and hire purchase companies and the acquisition and disposal of foreign securities and currencies.

Finally, item 12 (capital transfers) records the receipt of grants for designated capital expenditure from the European Regional Development Fund and the Agricultural Guidance and Guarantee Fund.

This brief description of the balance of payments capital account should clarify the interpretation of capital transactions and the distinction between debit and credit items. Estimates of the value of capital transactions are derived principally through the operations of public authorities, the Central Bank, the commercial banks and insurance companies, and some other financial institutions. The sum of all debit and credit items gives rise to a surplus or deficit on capital account. From Table 5.1 it can be seen that in 1978 there was a recorded surplus of £148.8m on capital account — inward capital movements exceeded outward movements. This surplus was matched by an equal and opposite deficit of £148.8m on current account. Taking all current and capital items together, total credits or inward movements equal total debits or outward movements — hence the surplus on capital account must be balanced by an equal deficit on current account.

The equality between total credits and total debits is a necessary consequence of the structure of the balance of payments accounts. As a simple illustration, suppose there were no autonomous capital movements, and that total current imports exceeded total current exports by £100m. To finance these extra imports, £100m of foreign exchange must be acquired, either by borrowing from abroad, or by running down external assets. In either case this accommodating capital transaction will appear as a credit item on capital account. Thus, in the balance of payments there will be a surplus of £100m on capital account and a deficit of £100m on current account, and total credits equal total debits.

Any imbalance between total autonomous current and capital debits and credits must be offset by an accommodating capital inflow or outflow. The demand for foreign exchange, arising from imports of goods and services, lending abroad and the purchase of external assets, must equal the supply, which arises from exports of goods and services, borrowing from abroad and the running down of external assets. Suppose, in addition to the £100m deficit on current account, the Central Bank wished to increase its net external funds by £50m. Taking both items together, this implies a deficit of £150m, and this excess demand for foreign exchange must be financed by an accommodating reduction in the external funds of the commercial banks or private investors, or by borrowing from abroad. This will result in a credit or inward movement of £150m on capital account — total credits and total debits will be in balance, and the surplus on capital account (£100m) will equal the deficit of £100m on current account.

Returning now to Table 5.1, the record of all capital transactions gives rise to a surplus of £148.8m. It follows that once all current transactions are recorded there should be an equal deficit on current account. However, since most of the transactions recorded are estimates which are subject to some margin of error, it would be surprising if the apparent deficit on current account were exactly equal to the surplus on capital account. The inevitable discrepancy between estimates of total credits and total debits is summarised in the item 'balance unaccounted for'. In Ireland, the practice is to treat such discrepancies as current in nature, so that the balance unaccounted for is entered as a credit or debit item on current account. One could assume that discrepancies are capital in nature, and include the balance unaccounted as a capital transaction. In practice the balance unaccounted for will include errors in the estimates for current and capital transactions.[14]

The structure and content of the balance of payments, and the trends in its constituents, are of considerable importance in relation to economic policy. Of immediate interest is the relation between total current imports and total current exports, and the trends in both merchandise and non-merchandise trade. Continuing deficits on current account simply mean that the proportion of current domestic output exported is insufficient to finance the current level of imports. In such circumstances the government may apply corrective policies designed to increase exports or reduce imports, or both, but these policy decisions are also influenced by trends in the external capital account. Persistent deficits which are financed by running down external reserves can clearly be maintained only for relatively short periods, depending on the actual level of external reserves. Alternatively, deficits may be financed by accommodating government loans raised abroad, though this is normally a fairly short-term measure usually accompanied by domestic policies designed to reduce or eliminate current balance of payments deficits. However, deficits which are financed by a continuing autonomous net inflow of private capital can be maintained for fairly long periods. This situation is characteristic of a developing country, which relies upon a substantial net inflow of capital to pay for its imports of capital goods and raw materials and to finance its developing industries. Eventually, with the development of domestic export industries, exports will be sufficient to finance current imports.

The experience of Ireland from about 1960 can be viewed against this pattern of development. High growth rates throughout much of this period resulted in persistent deficits in the current balance of payments, due to rising imports of raw materials and capital goods (as well as of consumption goods generated by higher personal incomes). These deficits were largely financed by the net inflow of investment funds, which helped to finance the development of export industries. In the absence of this net inflow of capital, more restrictive domestic economic policies would have been required to ensure equilibrium in the balance of payments and/or to increase the level of domestic savings.

Analysis and discussion of policy issues relating to Ireland's trade and balance of payments are extensively covered in the quarterly and annual reports of the Central Bank of Ireland. These include details of official external reserves, the external debt of the government and state-sponsored bodies, the external assets and liabilities of the banks, the volume and value of external trade (derived from the monthly trade statistics), and developments in exchange rates. The text of the reports includes regular analyses of internal and external monetary developments, balance of payments and exchange rate policies and development within the European Monetary System (EMS). Ireland joined the EMS in December 1978, simultaneously discontinuing the parity link with the pound sterling.

Notes to Chapter 5

[1]   3-year average 1976-1978. The share of imports in GNP varies markedly from year to year but has shown a consistent upwards trend.
[2]   For the period 1904-1921 some trade statistics for Ireland were published by the Department of Agriculture and Technical Instruction for Ireland.
[3]   These regulations are described in the introduction to the monthly trade statistics.
[4]   If a country has a surplus on invisible earnings (due perhaps to a thriving tourist industry) it can afford to import more merchandise than it exports. From the inhabitants' point of view, this is an advantage, though the expression 'adverse' balance of trade does not suggest that this is so.
[5]   These are: Live animals, food and food preparations; Beverages and tobacco; Raw materials except fuels; Mineral fuels and related materials; Animal and vegetable oils and fats; Chemicals; Manufactured goods classified by material; Machinery and transport equipment; Miscellaneous manufactured articles; Commodities and transactions not classified elsewhere.
[6]   Thus section 0 (Food and Live Animals) contains ten divisions numbered 00 to 09,

ranging from Live Animals chiefly for food (Division 00) to Miscellaneous Edible Products and Preparations (Division 09).

[7]   In some instances — for example, live animals — items are further sub-divided to a 6-digit level, while some other commodity groups for which the levels of international trade are very low are recorded at 4-digit level of detail.

[8]   A finer breakdown of country trade can be obtained by reference to the table of trade by commodities. In addition, full country by commodity detail can be obtained from the CSO by request (and payment of a small fee).

[9]   Various other factors affect the monthly volume of trade, such as the number of calendar days per month and the number of working days per month. The influence of these factors may be eliminated by using a 'standard month' or by measuring the 'volume of trade per working day'.

[10] Seasonally-adjusted estimates of the value and volume of trade were first published in 1978. The method of adjustment is the well-known US Bureau of the Census 'X-11 Variant' seasonal adjustment programme. For a summary of this method see Chapter 2 and the references cited there.

[11] The validity of the procedure outlined in this paragraph is less obvious than this simplified reasoning may suggest. For one commodity it is certainly true that $P_{ot} \times Q_{ot} = V_{ot}$, where P_{0t}, Q_{ot} and V_{ot} are respectively the price, volume and value indexes for the commodity, so that the volume index may be indirectly calculated by 'deflating' the value index by the price index. For a group of commodities, however, the validity of this relationship depends upon the index number formulae used. For example:

$$\frac{\Sigma P t \; q_o}{\Sigma P o \; q_o} \times \frac{\Sigma P o \; q_t}{\Sigma P o \; q_o} \neq \frac{\Sigma P t \; q_t}{\Sigma P o \; q_o} \; , \text{ while } \quad \frac{\Sigma P t \; q_t}{\Sigma P o \; q_t} \times \frac{\Sigma P o \; q_t}{\Sigma P o \; q_o} = \frac{\Sigma P t \; q_t}{\Sigma P o \; q_o}$$

even though in both expressions the left hand side is the product of a price index and a volume index. It does not necessarily follow, therefore, that deflating $\Sigma p_t \, q_t$ by a price index P_{ot} will give $\Sigma p_o \, q_t$ in an algebraic sense. In the present context, however, this point may be ignored.

[12] The actual methods used by the CSO are technically more sophisticated than the above description suggests, but the essence of the methods used is as outlined here.

[13] This item does *not* include, however, receipts from the *sale* of assets held abroad — transactions of this kind are included in the capital account.

[14] Margins of error in the estimates for certain 'invisible' items have already been noted. In addition, estimates of the value of merchandise trade are also subject to margins of error, due to smuggling, erroneous valuations, etc. In many countries, including the UK, the residual error of balance unaccounted for is included as a capital item.

6 National Income and Expenditure

The collection and presentation of national income and expenditure data have been greatly influenced by economic theory, particularly Keynesian and post-Keynesian macro-economic theory. A primary aim in the collection of such statistics was, and still is, to determine the total income accruing to persons and other factors of production over a given time period, which in turn involves the estimation of the total value of all goods and services produced by the community. Increasing importance has been attached to the breakdown of income and expenditure aggregates into components such as consumption expenditure, savings and investment; the relative shares of wages, salaries and profits in total income; the proportion of income paid in taxation and the role of the public sector; and the relations between different economic sectors and income and expenditure categories. Reflecting these developments in economic theory, the national income and expenditure accounts have tended towards greater detail and emphasis on the interdependence of the different constituents of the national accounts. This has aided empirical work designed to test hypotheses of economic theory, and the construction of econometric models.

The first official estimates of national income and expenditure for Ireland were made for 1938 and published in the *White Paper on National Income and Expenditure 1938-1944*.[1] A second White Paper covered the years 1945-1950,[2] and from 1951 to 1958 the estimates were published annually in the *Irish Statistical Survey*. From 1959, the national income estimates have been published separately in the *National Income and Expenditure* book. This annual publication includes detailed notes and explanations relating to the estimates, and it is important to read these carefully. Over the years frequent changes have been made in methods of estimation, and in the definition and breakdown of individual constituents of the accounts, so estimates for successive years may not be strictly comparable. Figures for earlier years are revised —

sometimes substantially — in the light of changes in sources or methods of estimation, but it is not always possible to do this in much detail, so that comparable figures over longer periods of years are available only for a certain number of aggregates. Series of estimates for most of the major items, however, are available on a comparable basis from 1947 onwards. As a result of the tendency for frequent revisions of the estimates, it is advisable to refer to the latest available National Income and Expenditure book, which contains the most up-to-date estimates for earlier years.

National income

The concept and measurement of national income may be approached in three ways — as income, as production, and as expenditure. The national income book defines national income as 'the total of all payments for productive services accruing to the permanent residents of this country'. This should be interpreted carefully, since not all forms of income accrue to persons, and national income is taken to include certain non-monetary and imputed items of income as well as actual monetary receipts and payments. To employ a lengthier definition, national income is the sum of all payments and returns (actual and imputed) to resident factors of production for current productive services.

Table 6.1 reproduces the official estimates of national income for the period 1975-78. Taken in conjunction with the official explanatory notes to the published tables, the general framework of this set of accounts is quite straightforward; national income or net national product at factor cost is estimated as the sum of all payments and returns to resident factors of production, and these are grouped in the accounts into different categories of income. Methods of estimation of the constituent items, and the accuracy of the estimates, are discussed later; here attention is concentrated on some basic concepts and definitions, on certain special features of the accounts and on the derivation of some important aggregates such as personal income and savings.

Items 1-3 represent incomes arising in agriculture, forestry and fishing. Included in income from self-employment is the estimated value of farm produce and fuel produced and consumed by farm households without process of sale (consumption of own produce), and the value of the changes in livestock numbers on farms, all valued at

Table 6.1 National income or net national product at factor cost 1975-78 (£m)

Category	1975	1976	1977	1978
Income from Agriculture, Forestry, Fishing:				
1. Income from self-employment and other trading income	486.3	549.8	758.9	862
2. Wages and salaries	42.6	46.2	50.6	55
3. Employers' contribution to social insurance	2.7	3.2	4.0	4
Non-Agricultural Income:				
Profits, professional earnings, interest, dividends and income from lands and buildings:				
Domestic				
4. Trading profits of companies (including corporate bodies) before tax	353.0	502.2	625.5	749
5. Income of Post Office and Post Office Savings Bank	12.7	17.5	27.2	46
6. Other trading profits, professional earnings, etc.	190.9	218.4	269.5	329
7. Adjustment for stock appreciation	—127.3	—189.6	—152.6	—112
8. Rent of dwellings (actual and imputed)	54.8	59.8	71.3	80
9. Rent element in land annuities	3.0	3.0	3.0	3
Remuneration of employees:				
Domestic				
10. Wages, salaries, pensions	1,970.9	2,330.3	2,718.5	3,190
11. Employers' contribution to social insurance	95.5	127.7	153.0	181
12. Adjustment for financial services	—92.8	—115.1	—160.8	—184
13. Net domestic product at factor cost	2,992.3	3,553.4	4,368.1	5,203
14. Net factor income from the rest of the world	—21.0	—2.1	—21.8	—26
15. Net national product at factor cost = National Income	3,013.3	3,551.3	4,346.3	5,177

Source: *National Income and Expenditure 1978.*

current agricultural prices. Item 1 therefore incorporates a significant non-monetary element of income. In estimating income from self-employment, which is essentially derived as the difference between the value of agricultural output and the costs of producing that output (see Chapter 3), note that rates are treated as an expense; hence Item 1 is net of rates.

Employers' contributions to social insurance (Item 3, and for non-agricultural employees, Item 11) are regarded as part of employees' income, albeit paid direct to government as a form of direct taxation. The incomes recorded in Table 6.1 are inclusive of income taxes and other direct taxes (i.e. are gross incomes), but are net of taxes on expenditure (rates, excise duties, VAT) and subsidies.

Items 4-12 refer to incomes arising from non-agricultural activities in Ireland. Trading profits are defined to include direct taxes but are net of indirect taxes (and subsidies). Since they refer to profits from economic activity within the state, they include the profits of Irish branches of foreign companies but exclude the profits of foreign branches of Irish companies.

In computing trading profits, no deduction is made for the payment of interest, dividends or rent; similarly, to avoid double counting, no allowance is made for the receipts of interest, dividends and rent. Such receipts and payments are simply transfers of factor income between different sectors and agents in the economy, and do not affect the aggregate value of factor earnings. For the same reason, the remuneration of employees and the income of farmers does not include the receipts of interest or other investment income.

An exception to this treatment is made for banks and similar financial institutions whose income is derived from the difference between interest paid to depositors and received from borrowers. If banks were treated in the manner described above, they would be shown to have large negative trading profits, since apart from investment income their only other revenue is from bank charges and commission, and these would not cover more than a small proportion of operating expenses. In the case of financial institutions, trading profits are defined to include the difference between interest and dividends received and interest paid to depositors, this difference being regarded as payment for the financial services rendered.

This treatment of the financial sector introduces an element of double counting into the national income estimates. The interest received by the financial sector is included in the trading profits of the companies (or persons) who pay the interest and part of it is also included in the financial sector's trading profits for national income purposes.[3] To compensate for this the national accounts include an item called 'adjustment for financial services', which is a negative amount corresponding to the difference between interest and dividends received by the banks and similar institutions, and interest paid to depositors.[4] This appears in Table 6.1 as Item 12. This treatment avoids the anomaly of the financial sector invariably showing large trading losses, and is consistent with the system of national accounts recommended by the United Nations (United Nations (1968)). For description of an alternative treatment, see *National Income and Expenditure 1976.*

Trading profits include the value of changes in stocks of finished goods, raw materials and work in progress between the beginning and end of the year. This change is a compound of changes in the volume or quantity of stocks and changes in their prices. That part due to price changes is not regarded as representing a 'real' change in output or value added, so an adjustment is made to aggregate trading profits to eliminate the effect of price changes on the value of net changes in stocks and work in progress. In periods of rising prices the adjustment will be negative, and in periods of falling prices it will be positive. The adjustment for stock appreciation for the non-agricultural sector is shown as Item 7 in Table 6.1. No adjustment is required for the agricultural sector, since the change in the value of livestock numbers is calculated directly on the basis of average annual prices.

Item 8 includes not only actual income received by property owners for rented accommodation but also estimated income enjoyed by owner-occupiers. In the case of local authority dwellings, the figures are based on rents received plus the value of housing subsidies paid by the local authorities. The estimates are net of depreciation, rates, repairs and maintenance.

Item 10 represents the estimated value of employee earnings in all sectors except agriculture, and is the largest component of national income. This figure is the sum of all wages, salaries, pensions and other earnings (overtime payments, bonuses, commission earnings, directors' fees, etc.) arising from economic activity within the state, and includes income in kind (food, fuel, shelter, clothing, etc.). It also includes employees' contributions to social insurance and contributory pension schemes and, where pension schemes apply, employers' contributions to the pension fund. Where pension funds do not exist, the actual pensions paid are included, the payments being regarded as superannuation contributions.

The income estimates in Table 6.1 represent payments for productive services; the latter including, for example, the activities of the armed forces, civil servants and similar 'non-marketable' activities. All such incomes may be classed generally as 'factor incomes'. Certain other types of income received by persons do not represent payments for which there is a corresponding current productive service; the most important are those which arise through the operation of various state social security services, including unemployment assistance and unemployment benefit, widows' and orphans' pensions, disability benefits, and various other types of pensions and allowances. Incomes of this kind are called 'transfer payments'[5] and are not counted as constituent items in national income. Payments of this kind are effectively a transfer of income between one section of the community and another, and are already included as part of total factor income. Details of these transfer payments are included in the annual national income book. Interest payments on the national debt are also regarded as transfer payments and excluded from national income.[6] The value of domestic work by housewives is not included in national income, but the income in cash and kind of domestic servants is included. The definition of transfer incomes is by no means clear-cut, and the treatment of borderline cases is determined by convention and practical convenience. The principle underlying the treatment of transfer payments is the avoidance of double counting, i.e. to include in national income only those forms of income which represent payments for 'productive services'.

Depreciation

Part of incomes earned during the year must be set aside to replace capital equipment and plant used up in the production of goods and services. This is called the 'provision for depreciation'. Where applicable (e.g. trading profits) an estimate of depreciation is deducted from total factor incomes and the national income figures recorded in Table 6.1 are net of depreciation. The provision for depreciation, recorded in subsequent tables in the national income book, is supposed to represent the amount required to maintain the community's capital assets intact.

The estimated provision for depreciation in any year is a function both of the value of the community's stock of capital at the beginning of the year and the expected working lives of the assets which constitute this stock of capital. Since neither the value or composition of the country's stock of capital assets are known with any certainty, and the working lives of different assets are highly variable, the national accounts' estimates of depreciation are necessarily hypothetical figures and subject to a considerable degree of error. The estimate for depreciation is in any case hypothetical; it purports to show what sum is required to maintain capital, not what is actually spent.

In the absence of reliable data on the direct measurement of 'real' depreciation the estimates are based on a variety of sources and methods. Estimates at replacement cost are made for the agricultural sector, based on data collected in the annual agricultural enumeration (see Chapter 3). Replacement cost estimates are also compiled for housing. For the other market sectors of the economy, estimates are based on allowances for wear and tear for tax purposes and other tax allowances. Consequently, these estimates of depreciation are affected more by fiscal considerations than by the actual rate of wear and tear and obsolescence, though it is likely that these estimates are now more in accord with 'real' depreciation costs than they used to be.

The figure for depreciation published in the national income book covers depreciation in agriculture, industry, dwellings and buildings owned by public authorities. Estimates of depreciation are somewhat hazardous so the published estimates of national income, which are net of depreciation, should be regarded with caution. The constituent elements of national income are also subject to errors of estimation of varying degrees, so that minor variations in the level of national income cannot be regarded as significant.

The sum of items 1-12 in Table 6.1 is called *net domestic product at factor cost*. It is 'net' because of the deduction of the provision for depreciation discussed above. It is 'at factor cost' because indirect taxes and subsidies are excluded. It is 'domestic' because it refers only to incomes generated from economic activity within the state. Profits from the operations of Irish concerns abroad are not included in domestic trading profits, while the profits of Irish branches of foreign companies are included. Similarly, the dividends and interest received from investment abroad by Irish residents are not included in domestic income, while dividends and interest paid abroad are included. Domestic income or product thus measures the aggregate value of incomes generated by economic activity within the state, whether or not all that income is actually received by Irish residents, and excludes payments received as a result of economic activities outside the state.

If the outflow of dividends and interest to foreign investors and of the profits (less taxes paid to the state) of Irish branches of foreign companies are deducted from net domestic product at factor cost, and the inflow of dividends and interest and the profits of foreign branches of Irish companies are added, we arrive at net *national* product at factor cost, or *national income*, i.e. the sum of incomes received by residents of the state.

These adjustments are shown in Table 6.1. as item 14, net factor income from the rest of the world. This item also includes the net remuneration of employees from the rest of the world — on the credit side, remuneration received by Irish residents from non-resident employers, and on the debit side payments made to non-resident employees by Irish concerns.

The figure recorded is the net balance of these factor incomes. A positive value means that the inflow of factor incomes exceeded the outflow, while a negative value means that the inflow was less than the outflow. From Table 6.1 it can be seen that the net balance was positive in 1975, but negative thereafter.

The only constituents of national income which do not ultimately accrue to persons are the trading income of public authorities[7] and the undistributed profits of business enterprises. By convention, it is assumed that all the profits of unincorporated enterprises are distributed and therefore form part of personal incomes. It is more convenient to do this, and to count the retained profits of unincorporated enterprises as part of personal savings, than to attempt to estimate directly the proportion of incomes of this kind which are 'ploughed back' into the business. If, then, the undistributed profits of companies and the trading income of public authorities are deducted from national income the remainder will represent total factor incomes received by persons. The addition to this of transfer incomes of the type mentioned above gives *total personal income*. This figure includes income in kind, imputed income and the value of social insurance contributions, since these items are included as part of national income. Since personal income includes both factor incomes before taxation and transfer incomes, it is quite possible that it may exceed national income in value. Transfer payments are simply redistributed factor incomes, and there is therefore an element of double-counting in total personal income.

Personal income is paid in taxation, spent on consumer goods and services or saved. Consequently, if taxes on income (including social insurance contributions and taxes on personal wealth) and savings are deducted from personal income, the remainder will be a residual estimate of total personal expenditure on consumer goods and services. Alternatively, personal savings may be estimated as a residual by deducting from personal income the sum of taxes on income and personal expenditure. In either case, the residual will reflect any errors in the estimates for the other two items, as well as the estimate for the control total of total personal income. The figure for personal savings recorded in the accounts is estimated in this way, although the estimate for personal expenditure is also calculated as a residual on the expenditure side of the accounts, as will be explained later.

The national income accounts also include estimates of what is called 'private income'. This is national income less the trading and investment income of public authorities plus transfer payments, and differs from personal income in that it includes the undistributed profits of companies before tax. Of greater interest is total personal income and its distribution between taxation, savings and personal expenditure. 'Taxes on personal incomes' include social insurance contributions, taxes on dividends and interest received by persons and taxes on personal wealth, as well as the normal income tax levied on earned incomes.

Indirect taxes such as rates and other forms of taxes on expenditure are excluded. This should be remembered in discussing problems such as the proportion of incomes paid in tax. For some purposes interest would lie only in the amount and proportion of incomes paid in direct taxation; if, for example, it was desired to examine consumption/savings patterns, direct taxes would be deducted from personal income to give disposable income, from which ratios such as the marginal propensity to consume could be calculated. Here indirect taxation is of little interest. In a different context it may be of interest to examine all taxes paid by persons, including taxes on expenditure, and to compare the relative yields of direct and indirect taxation. Details of the yield from indirect taxation are recorded in other tables in the accounts; these include a wide variety of taxes, not all of which are levied on consumer goods and services.

Personal saving is arrived at as a residual by deducting from personal income the sum of direct taxes and personal expenditure on current goods and services. This includes a non-monetary element, notably the value of changes in livestock. Since the latter is included as part of the personal income of farmers, it will also be included as part of personal savings.

Table 6.2 shows the derivation of private and personal income and personal savings for the period 1975-78. The starting figure for national income includes stock appreciation, so the residual figure of personal savings will also include stock appreciation. National income *less* government trading and investment income *plus* transfer payments (including national debt interest) yields total private income. Private income *less* the undistributed profits of companies gives total personal income. The last three lines of the table show the distribution of personal income between expenditure on consumers' goods and services, taxes on income, and personal savings.

Table 6.2 Personal income and expenditure 1975-1978 (£m)

	1975	1976	1977	1978
National income before adjustment for stock appreciation	3,140.6	3,740.9	4,498.9	5,289
Less Government trading and investment income	−78.4	−93.8	−135.2	−171
Plus National debt interest	170.6	241.1	307.9	395
Plus Other transfer incomes	564.6	665.6	765.7	880
Private income	3,797.4	4,553.8	5,437.3	6,393
Less Undistributed profits of companies before tax	−153.0	−233.1	−309.8	−371
Personal income	3,643.5	4,320.7	5,127.5	6,022
Distributed between:				
Personal expenditure on consumers' goods and services	2,369.9	2,865.1	3,357.1	3,915
Taxes on personal income and wealth	485.0	663.2	803.6	908
Personal savings	788.6	792.4	966.8	1,199

Source: *National Income and Expenditure 1978*

The sum of personal savings and companies' savings is called total private savings. Companies' savings are calculated as the undistributed profits of companies, less actual tax payments on profits. The qualification 'actual' is necessary, since what is owed by companies at current tax rates is not necessarily paid to the revenue authorities in that year. The taxes paid by a company in the current year are based upon the company's income in the previous year; taxes on the current year's income are paid in the following year. If the company's annual income is rising, its current liability for tax will exceed actual current tax payments and this difference, familiar to shareholders and others who examine company balance sheets, is referred to as 'additions to tax reserves'. Since these tax reserves are not actually paid to the revenue authorities in the current year, they are available for investment and therefore form part of companies' savings. In times when trading profits are rising rapidly, additions to tax reserves may form a substantial element in company savings. If trading profits are falling, or if there is a reduction in tax rates, additions to tax reserves may be negative.

Public authorities account for the third source of savings. Adding this to total private savings gives total savings before adjustment for stock appreciation. Adjusting for stock appreciation gives total savings, which is the total amount available from domestic sources for net investment expenditure. Actual net domestic investment, however, will be affected by the flow of funds into and out of Ireland. A net capital inflow of funds will increase the amount available for domestic investment, and conversely for a net capital outflow. In Ireland there is almost invariably a net capital inflow, thus increasing the funds available for domestic investment. The net capital inflow is equal to the balance of payments current deficit. The net capital inflow from abroad, termed 'net foreign disinvestment' in the national income book, added to total domestic savings, represents the total funds available for net domestic capital formation. If the estimated provision for depreciation is added to this, we have the total amount available for gross domestic capital formation. The actual distribution of these investment funds is discussed later. The composition of national savings 1975-78 is shown in Table 6.3.

Table 6.3 The composition of total savings 1975-78 (£m)

	1975	1976	1977	1978
Personal savings	788.6	792.4	966.8	1,199
Companies' savings	96.2	161.7	232.0	266
Total private savings	884.8	954.1	1,198.8	1,465
Public authorities' savings	−244.6	−179.5	−193.6	−300
Total savings before adjustment for stock appreciation	640.2	774.6	1,005.2	1,165
Adjustment for stock appreciation	−127.3	−189.6	−152.6	−112
Total savings	512.9	585.0	852.6	1,053
Net foreign disinvestment	6.0	157.1	155.2	149
Provision for depreciation	298.1	372.0	468.6	585
Gross amount available for domestic physical capital formation	817.0	1,114.1	1,476.4	1,787

Source: *National Income and Expenditure 1978*

An alternative measure of national income

National income was discussed above from the distribution side, as the sum of payments and returns to resident factors of production. The measurement of national income may also be approached from the side of production, as the contribution by different sectors of the economy to the value of national income. The output of any sector — agriculture, industry, services — consists of the value of goods and services bought from other sectors plus the value added by the sector itself. The 'value added' by each sector is composed of the wages and salaries, profits and other factor payments generated within the sector, and represents the contribution of each sector to the total value of goods and services produced by that sector. Goods and services bought by one sector from other sectors (or imported) do not form part of that sector's contribution to the final value of output, since they are already included as part of the value added by other sectors. It follows that the annual value of all *final* goods and services produced is conceptually equivalent to national income, and provides an alternative means of measuring it.

This identity between income and output may be clarified by the following example:

	Farmer	*Miller*	*Baker*
Cost of materials	—	25	50
		(wheat)	(flour)
Value added	25	25	25
Value of output	25	50	75
	(wheat)	(flour)	(bread)

For convenience it is assumed that farmers do not purchase fertilisers or any other materials, so that the output of the farming sector consists only of value added, which is the farmers' income. Wheat is sold to the milling sector, which transforms it into flour and sells its output to the baking sector, the final stage in the production process. The national income of this simple economy may be calculated by adding together the value added by each sector (25 plus 25 plus 25 = 75), or as the total value of the output at the final stage of the production process (75, the value of the output of the baking sector). Given the definition of value added, a few moments' reflection will show the relation between value added, the value of output of final goods and services, and national income.

If all factors of production are grouped into a number of different sectors, and estimates are made of value added by each sector, the sum of value added by each sector should equal national income. Since every factor of production is included, and since value added is defined to include all payments to factors of production, the sum of value added by each sector will include all factor earnings, and should equal national income. This result can be obtained by re-classifying the income estimates in Table 6.1 to show income by sector of origin rather than by type of income. This is an alternative classification of national income. It is possible to start from the output side and to arrive at an independent estimate of national income by methods analogous to those illustrated in the example above. Although in Ireland output estimates do not appear to play a significant independent role in national income estimates,[8] in some countries output estimates provide a more reliable basis for the calculation of national income than alternative methods.[9] Output estimates can play an important role in describing flows of goods and services between different sectors of the economy and in helping to establish consistency between different sets of accounts and income and expenditure estimates (see pp. 141-6).

Data on income by sector of origin are shown in Table 6.4. Five sectors are distinguished — agriculture; industry; distribution, transport and communications; public administration and defence; and other domestic. Value added in each sector is allocated between remuneration of employees (labour income) and other (profits and rent). Starting with the estimates of value added by sector, several adjustments are required to arrive at national income. The estimates for value added include the effects of stock appreciation, except in agriculture where the value of physical changes in stocks is calculated directly. It is also necessary to add in the adjustment for financial services. The sum of value added in each sector, plus the adjustments referred to, gives net domestic product at factor cost, identical to that already obtained as the sum of factor incomes in Table 6.1. The addition of net factor income from the rest of the world yields the familiar net national product at factor cost = national income.

These aggregates are 'net' because the value added estimates for each sector recorded in the accounts are net of depreciation. If the provision for depreciation is added to both these aggregates, we obtain respectively *gross domestic product* and *gross national product* at factor cost.

The output of each sector is the sum of the value of goods and services bought from other sectors and the value added by the sector itself. This statement must be qualified to account for the effects of indirect taxes and subsidies. Taxes on output raise the selling value (market value) of products which are taxed, whilst subsidies have the opposite effect. Returning to the example quoted at the beginning of this section, if a tax were levied on bread, the market value of the output of the bakery sector would be greater than 75, and similar effects would be observed if the tax were levied at an earlier stage in the production process, such as

Table 6.4 National product at factor cost and at market prices 1975-78 (£m)

	1975	1976	1977	1978
Agricultural:				
Agriculture, forestry & fishing				
Remuneration of employees	45.3	49.4	54.6	59
Other	489.3	552.8	761.9	865
Non-Agricultural:				
Industry				
Remuneration of employees	856.1	1,008.9	1,201.8	1,424
Other	260.0	361.1	440.1	528
Distribution, transport and communication				
Remuneration of employees	400.0	475.6	552.5	648
Other	176.8	230.7	279.3	351
Public administration & defence				
Remuneration of employees	235.2	281.9	310.5	362
Other domestic (including rent)				
Remuneration of employees	575.1	691.6	806.7	937
Other	174.6	206.1	274.1	325
Adjustment for stock appreciation	—127.3	—189.6	—152.6	—112
Adjustment for financial services	—92.8	—115.1	—160.8	—184
Net domestic product at factor cost	2,992.3	3,553.4	4,368.1	5,203
Net factor income from the rest of the world	21.0	—2.1	—21.8	—26
Net national product at factor				
cost = national income	3,013.3	3,551.3	4,346.3	5,177
Provision for depreciation	298.1	372.0	468.6	585
Gross national product at factor cost	3,311.4	3,923.3	4,814.9	5,762
Taxes on expenditure	645.8	889.1	1,012.0	1,112
Subsidies	—256.2	—304.8	—487.4	—629
Gross national product at current				
market prices	3,701.0	4,507.6	5,339.5	6,245

Source: *National Income and Expenditure 1978.*

a tax on flour or wheat. If output estimates are taken to include these indirect taxes and subsidies, the corresponding aggregates obtained are called gross (or net) domestic product and gross (or net) national product *at market prices.*

Thus net national product at market prices differs from national income to the extent of the value of indirect taxes (positive) and subsidies (negative). If output estimates are net of indirect taxes and subsidies, the national product aggregates are said to be valued at factor cost. It is important to know which valuation is being used. In 1977, for example, GNP at factor cost was £4,814.9m and GNP at market prices

£5,339.5m (indirect taxes + £1,012m, subsidies – £487.4m); between 1977 and 1978 GNP at factor cost increased to £5,762m (an increase of 19.7 per cent), whilst GNP at market prices increased to £6,245m (an increase of 17.0 per cent). Rather different percentage changes are noted, and it is clearly important to know which set of figures is being used.

The wheat-flour-bread example quoted above can be used to show the combined effect of subsidies and indirect taxes on the calculation of national product at factor cost and at market prices. Suppose a subsidy of 5 units of value is paid to farmers and a value added tax of 20 per cent is levied on flour producers and on bread producers. The amended table would be as follows.

		Farmer	Miller	Baker
1	Cost of materials	—	20	50
			(wheat)	(flour)
2	Value added	25	25	25
3	Subsidy	-5	—	—
4	VAT	—	5	5
5	Value of output at market prices	20	50	80

The national product at factor cost or national income is obtained as before as the sum of value added in each sector (line 2), giving 75, or as the total value of output at the final stage of production *less* indirect taxes (80 — 5 = 75). National product at market prices is equal to national income *plus* total indirect taxes *less* subsidies, i.e. 75 + 5 + 5—5 = 80, or as the total value of output at the final stage of production (80), inclusive of indirect taxes and/or subsidies at that stage.

Estimates of net national product by sector of origin, and gross national product at factor cost and at market prices, are reproduced in Table 6.4. The figure for net national product by sector of origin is a re-classification of national income recorded in Table 6.1. All types of imputed income and income in kind are included under the appropriate sector heading. In relation to the sector 'public administration and defence' it should be noted that certain trading activities of local and central government (e.g. the post office), as well as health and educational services, are included as part of the output of other sectors. The 'output' of the public administration sector consists of the wages and salaries paid to employees of central and local government who are engaged in an administrative or regulatory capacity, including the income in cash and kind of armed forces personnel. In most cases the services of public authorities are not sold commercially but provided 'free'; for the purpose of national income accounting, it is assumed that the value of the output of public authorities' services is what it cost to produce them. The contribution of public authorities to national product consists of the total cost of public authorities' services *less* the cost of goods and services bought from other sectors — that is, value added.

Measures of expenditure

Income is generated by the production of goods and services. In the previous section it was shown that national income = net national product at factor cost. A third way of measuring national income is to estimate expenditure on goods and services. Total expenditure on goods and services must equal gross (or net) national product (the total value of goods and services produced) and national income (the total of incomes generated in producing these goods and services). Measured from the output side, national product is the value of all goods and services produced according to sector of origin; measured from the expenditure side, national product is the value of all goods and services produced according to final use — that is, for consumption, for investment, for export.

Since stock changes are included as part of investment expenditure, the total annual outlay or expenditure on goods and services must equal the total value of goods and services produced. Income, output and expenditure are simply three different ways of evaluating the same aggregate.[10] Referring back to the example shown at the beginning of the previous section, the total expenditure on final goods and services is 75 — i.e. consumption of bread. Expenditure on intermediate products, like wheat and flour, is not counted, since the value of these products is already included as part of expenditure on bread.

The national product can be measured 'gross' or 'net', and at factor cost or market prices. Similarly national expenditure can be measured gross or net, and at factor cost or market prices. Since expenditure on goods and services is normally expressed at market prices, it is convenient to start by calculating gross national expenditure at market prices. Other aggregates may then be derived from this by making suitable adjustments for depreciation and indirect taxes and subsidies.

The relation between the various income, output and expenditure aggregates is shown in Table 6.5. The expenditure categories listed in lines 1-6 represent the allocation of gross domestic product amongst different uses or forms of final demand. These categories (personal consumption expenditure, investment, etc) reflect a taxonomy founded on Keynesian theories of income determination, and are universally used in national income accounting.

Each of the first four categories listed includes expenditure on imported goods and services, but excludes that part of domestic output which is exported (for the purposes of national income accounting all exports are classed as final goods since they undergo no further transformation within the country). The sum of the first four items gives the total supply of goods and services available for domestic use. Subtracting imports and adding exports gives the total domestic output of final goods and services or gross domestic product at market prices. Gross domestic and gross national product differ to the extent of net factor income flows between Ireland and the rest of the world. If these are added we get gross national product at market prices. Each expenditure category in the table is valued at current market prices, inclusive of indirect taxes and subsidies. If indirect taxes are deducted from GNP at current market prices, and subsidies added, the result is GNP at factor cost.

Item 3 — gross domestic fixed capital formation — measures the total outlay on capital goods. Part of this expenditure is needed to replace plant and equipment used up during the year, and corresponds to the provision for depreciation. The estimated provision for depreciation is subtracted from gross national product at factor cost to give net national product at factor cost = national income.

Since income and expenditure are equal, independent estimates of expenditure on gross national product may be compared with alternative estimates of income or output, and the results used to check the accuracy of the different sets of estimates, or to indicate likely sources of error. In practice, expenditure estimates do not play a significant role in reconciling alternative estimates of GNP in Ireland. Personal expenditure (item 1) is regarded as a residual estimate and fixed so that the income and expenditure sides of the accounts balance. This should not be regarded as a Machiavellian deception by the CSO. Since income and expenditure are equal, any disparities between the income and expenditure estimates must be due to errors in the estimates for one or more of the constituent items. In Ireland, GNP is

Table 6.5 Expenditure on gross national product 1975-78 (£m)

	1975	1976	1977	1978
1. Personal expenditure on consumers' goods and services	2,369.9	2,865.1	3,357.1	3,915
2. Net expenditure by public authorities on current goods and services	688.1	844.7	980.8	1,130
3. Gross domestic fixed capital formation	870.3	1,101.1	1,367.3	1,769
4. Value of physical changes in stock	—53.3	+ 13.0	+ 109.1	+ 18
5. *Plus* exports of goods and services	1,619.0	2,153.0	2,803.8	3,369
6. *Less* imports of goods and services	—1,814.0	—2,467.2	—3,256.8	—3,930
7. Gross domestic product at market prices	3,680.0	4,509.7	5,361.3	6,271
8. Net factor incomes from abroad	21.0	—2.1	—21.8	—26
9. Gross national product at market prices	3,701.0	4,507.6	5,339.5	6,245
10. *Less* taxes on expenditure	—645.8	—889.1	—1,012.0	—1,112
11. *Plus* subsidies	256.2	304.8	487.4	629
12. Gross national product at factor cost	3,311.4	3,923.3	4,814.9	5,762
13. *Less* depreciation	—298.1	—372.0	—468.6	—585
14. Net national product at factor cost = national income	3,013.3	3,551.3	4,346.3	5,177

Source: *National Income and Expenditure 1978.*

estimated from the income side; after the estimates for the other expenditure categories have been made, the estimate for personal expenditure is adjusted to bring expenditure into line with income. This assumes that the income estimates are 'correct' and that any errors in the estimates can be attributed to the estimate for personal expenditure. This method of balancing the accounts is further discussed below.[11]

Estimates for the other categories of expenditure in Table 6.5 are based on a variety of sources. The methods of compilation of exports and imports have already been explained. The difference between imports and exports is the balance of payments on current account.

It is convenient to divide the remaining items into current and capital expenditure. Capital expenditure is the sum of items 3 and 4. Item 4 includes the familiar 'adjustment for stock appreciation', i.e. it is less (plus) that part of the change in value due to a rise (fall) in the prices of opening stocks. This treatment ensures consistency with the income and production accounts. Item 3 — gross domestic fixed capital formation — consists of total expenditure, both public and private, on building and construction, machinery and other capital goods such as commercial vehicles. With the exception of private dwellings, it excludes expenditure on consumer durable goods, such as motor vehicles for private use, washing machines, television sets, refrigerators.[12] By convention, such products are regarded as consumed in the year in which they are bought and are included in the personal expenditure category. This results in an understatement of the annual amount by which the community is adding to its stock of capital assets. This treatment of consumer durable goods is determined by practical considerations; if consumer durables were to be treated as capital assets, it would be logical to count the value of the services which they provide as imputed income to their owners, over the life span of the assets. But, except in the case of private dwellings, it would be virtually impossible to determine the value of this imputed income stream. On the other hand, gross domestic fixed capital formation includes various other types of capital expenditure, such as expenditure on churches, public parks and similar items, which also do not yield any readily measured flow of goods or services in future years. In principle, there is no reason why expenditure on consumer durables should not be similarly included as part of capital formation. Most other countries also treat consumer durable expenditure as part of personal current expenditure; in some countries, however, expenditure on certain consumer durables, such as private cars, is included with capital expenditure. This is another factor to remember in making international comparisons.

Capital formation may be estimated 'gross' or 'net'. Net capital formation is the more informative figure, since it measures the change in the stock of capital assets: it is regarded as an important determinant of the rate of growth of GNP. Unfortunately it is difficult to calculate with any degree of accuracy since depreciation, or capital consumption, is notoriously hard to assess. Figures for capital formation are therefore usually quoted gross.

The sum of items 3 and 4 in Table 6.5 constitutes gross domestic physical capital formation. The relative share of capital formation in

GNP measures the country's sacrifice of present consumption for future consumption, by adding to the stock of capital assets and thereby increasing the prospective flow of goods and services in the future. It is assumed that the higher the current rate of (net) capital formation, the higher will be the future rate of growth of GNP. However, while a high rate of capital formation may be taken as a necessary prerequisite for a high rate of growth of GNP, the rate of growth will also depend upon the distribution of investment funds amongst different uses (as well as upon the utilisation of capital). A distinction is often made between 'productive' and 'non-productive' investment; the former refers to investment in the industrial and agricultural sectors, while the latter includes 'social' investment such as housing, hospitals and schools. It is unwise to draw too hard and fast a distinction between different types of investment expenditure on this basis.[13] It is certainly the case that some forms of investment will be more conducive to growth than others; of particular importance is the proportion of investment devoted to plant and equipment for manufacturing industry.

Some details of the composition of gross capital formation are given in the national income accounts. One table identifies capital expenditure on dwellings, roads, other building and construction, transport equipment, agricultural machinery, other machinery and equipment, changes in livestock numbers and changes in non-agricultural stocks. Another distinguishes home-produced from imported capital goods, and a further table presents a breakdown of public authorities' capital expenditure (mainly building and construction). The national income book does not include a breakdown of capital formation by industrial sector, which would be valuable for analyses of growth in different sectors of the economy and the estimation of sectoral incremental capital-output ratios,[14] though such a breakdown is compiled and published along with comparable data for other member countries by the Statistical Office of the EEC.

Investment is financed by savings. Except in the case of the undistributed profits of companies, savings and investment decisions are normally independent of one another, but it is a consequence of the definitions of income and expenditure used for national income accounting that total savings = total investment expenditure. Investment expenditure is that part of domestic expenditure which is not spent on goods and services for current consumption. On the income side, what is not spent as current consumption of goods and services constitutes savings. Since income and expenditure are, by definition, equal, savings = investment.[15] This formal equality, of fundamental importance in the theory of income determination, plays an important role in the construction and integration of the national income and expenditure accounts.[16]

The principal constituents of domestic savings are personal savings, the undistributed profits of companies and the savings of public authorities. Adding to these the provision for depreciation, the sum of these four items, adjusted for stock appreciation, represents the total funds available from domestic sources for investment expenditure. But domestic sources may be supplemented by borrowing from abroad; conversely they may be reduced by lending abroad. This net foreign borrowing or lending is the familiar balance of payments on current account, which appears on the expenditure side of the accounts as the difference between current exports and current imports. In Ireland there is almost invariably a deficit on current account, involving net borrowing from abroad (termed 'net foreign disinvestment' in the accounts) and an increase in the funds available for domestic investment expenditure. In the last few years net foreign borrowing has played a very important role in the finance of investment in Ireland, and is likely to continue to do so. In a period of relatively rapid growth, investment expenditure has outstripped domestic savings, and the gap has been filled by borrowing from abroad.[17]

Total domestic expenditure is shown in Table 6.5 as the sum of items 1-4, in 1978 £6,832m at current market prices. What was earned or produced in 1978 was £6,271m at current market prices (item 7). This difference of £561m was partly financed by the net inflow of investment income, remittances and other transfers from abroad, amounting in 1978 to £412m. The remainder (£149m) was financed by net capital inflow or net foreign borrowing. (These last two figures are derived from data on Ireland's international balance of payments which are also recorded in the national income book).

Net foreign borrowing, sometimes on a substantial scale, is a necessary condition for growth in a developing economy. A principal characteristic of developing countries is a low level of domestic savings, so that a substantial net inflow of capital is a valuable source of finance for investment. Here net foreign borrowing provides a supplement to an inadequate rate of domestic savings; in other cases net foreign borrowing may be regarded more as a substitute for domestic savings, in situations of domestic inflationary pressure and high levels of personal consumption expenditure.

Details such as the share of foreign borrowing in total available investment funds, the distribution of personal income between saving and consumption, the rate of growth of personal consumption and the level and composition of capital expenditure are important indicators of trends in the economy. In Ireland's case, while the rate of domestic savings is comparable with that of other developed countries, the process of industrialisation has led to a rate of investment which has exceeded the supply of domestic savings (in recent years the gap has been exacerbated by large public sector borrowing requirements) and foreign borrowing has helped to sustain the level of investment.

Net foreign borrowing, the provision for depreciation and the savings of public authorities are estimated from a variety of sources. The savings of companies are also estimated, as the undistributed profits of companies less tax payments on these profits. Additions to tax reserves are included with company savings. The remaining constituent of total savings is personal savings, and this is obtained as a residual estimate by deducting personal expenditure and direct taxes from total personal income. As a result of the structure of the national accounts, personal savings appear as the difference between gross domestic fixed capital formation and the other known elements in total savings. Being a residual estimate, this figure will reflect errors in the estimates for other items. For instance, if capital formation is over-estimated then the estimate for personal expenditure on consumer goods and services will be under-estimated; this will mean that personal savings will be over-estimated by the same amount.

The level of personal savings will also be affected by definitions of gross capital formation. As one example, changes in the value of livestock on farms are included as part of gross capital formation. Consequently, changes in the value of livestock must also be included as part of personal income and personal savings.

The principal constituent of current expenditure is personal expenditure on consumers' goods and services, including expenditure on consumer durables. Personal expenditure is calculated by deducting the other expenditure items from the estimated value of GNP at market prices, and will therefore reflect any errors in these other estimates. Details of personal expenditure are recorded in later tables in the national income accounts, in which aggregate expenditure is broken down into about fifteen different categories, including household consumer durable goods and motor vehicles for private use. These individual expenditure categories are directly estimated on the basis of census of production data and import statistics, and then adjusted so that their total agrees with the control total of aggregate personal expenditure.

The individual expenditure categories recorded in the accounts are of interest in indicating general trends in the level and pattern of consumption expenditure. In interpreting the figures, several points should be borne in mind. The estimates are inclusive of the effects of indirect taxes and subsidies, changes in which will be reflected in year-to-year movements in individual categories of expenditure. Secondly, the figures include expenditure on imported consumer goods and services. Thirdly, they include expenditure by tourists and other visitors to Ireland (see Chapter 7 below). Before 1956, a conjectural adjustment was made to each expenditure category to allow for expenditure by tourists; since 1956 each expenditure category has included expenditure by tourists and a global adjustment is made to total personal expenditure. (The estimate for total tourist expenditure is then transferred to 'exports'.) The reason for this change of method was the difficulty of accurately estimating the breakdown of tourist expenditure. Finally, the expenditure categories include certain items of imputed expenditure — imputed rent of owner-occupied dwellings, consumption of own produce by farmers, expenditure on 'free' food, accommodation, clothing, etc. by domestic servants, agricultural employees and armed forces personnel. These items must be included since they are counted as part of personal incomes.

Public authorities

The remaining item of current expenditure in Table 6.5 is 'net expenditure by public authorities'. Details of the income and expenditure of local and central government, and the combined accounts of public authorities, are recorded in the national income book. The structure and purpose of public authorities' accounts are somewhat different from those of the national income and expenditure accounts, and their relation to one another requires careful interpretation. The income of public authorities consists of various forms of tax revenue — taxes on income, taxes on expenditure (including rates) and taxes on capital — and other income arising from the trading and investment activities of public authorities. Other items of revenue, e.g. current grants from central government to local authorities and interest payments on national debt held by local authorities, are transfer payments between public authorities and are excluded from the combined accounts of public authorities. Public authority expenditure consists of subsidies, transfer payments and current expenditure on goods and services; the latter includes wages and salaries paid to employees of public authorities and payments to persons and companies outside the public sector for goods and services. Current expenditure in the form of payments between public authorities — e.g. current grants from central to local authorities — are excluded from the combined public authorities' accounts.

That part of public authorities' expenditure which falls under the headings of subsidies or transfer payments is not included in the public authorities' share of GNP. Transfer payments and subsidies do not represent expenditure on gross national product by public authorities; they are money transfers which are included in the income and expenditure of persons who receive them. In calculating the public authorities' expenditure on gross national product, therefore, we are left with the current expenditure of public authorities on current goods and services, the only part of their expenditure which represents actual outlay on gross national product.

The major share of public authorities' expenditure on current goods and services is accounted for by wages and salaries of employees. The balance consists of the purchase of current goods and services from outside the public sector. For the purpose of the national income accounts, further adjustments are required. The item 'net expenditure by public authorities on current goods and services' (item 2 in Table 6.5) is not simply the sum of current expenditure on goods and services by central and local government which is recorded in the combined accounts of public authorities. For example, in 1978 current expenditure by public authorities on goods and services was £1,265m. Adjusted for inclusion as part of expenditure on gross national product, however, this figure was reduced to £1,130m, described as 'net' current expenditure by public authorities. The difference of £135m was accounted for by various items of miscellaneous receipts on the income side of the public authorities' accounts. The purpose of this adjustment is to avoid double-counting of expenditure. For example, local authority services for which a fee is charged will be included as part of personal expenditure on current goods and services, and their cost must therefore be excluded from local authorities' outlay on gross national product, or they would be counted twice. Most public authority services are provided 'free', so the cost of providing such services is regarded as expenditure by public authorities. In addition certain appropriations-in-aid and extra-budgetary funds, for which there is no corresponding reduction in the current income (and hence expenditure) of the private sector, are also deducted. The deduction of these miscellaneous receipts gives the figure of net current expenditure by public authorities which is included as part of outlay on gross national product.

The previous paragraphs have dealt with the current income and expenditure of public authorities. Capital expenditure by public authorities is included as part of gross fixed domestic capital formation (item 3, Table 6.5). A breakdown of public sector capital expenditure is recorded in a later table in the national income accounts. Remarks similar to those made above in relation to the public authorities' current account apply to the interpretation of the capital account. The contribution of public authorities to total gross physical capital formation consists of direct investment expenditure by local and central government (roads, housing, drainage, post office construction work, etc.); capital grants and loans to persons or enterprises are transfer payments for which the corresponding expenditure appears as part of private investment expenditure.

GNP at constant prices

A common use of national income statistics is to examine changes in the standard of living. The 'standard of living' or 'welfare' of a country cannot be precisely defined, but there is some relation between changes in income and changes in the standard of living. In using the national income accounts for this purpose there are several points to remember.

The estimates for gross national product, national income and the principal constituents of GNP published in the accounts are quoted at current money values. As a result of changes in prices, gross national product may increase from one year to the next without a corresponding increase in the real volume of goods and services available. In these circumstances it could not be claimed that there had been an increase in real national product or the standard of living. To measure the real change in national product it is necessary to estimate the value of output or expenditure at constant market prices. One method, used by the CSO, is to re-value the constituents of expenditure on gross national product at constant market prices, the present 'base' year for these estimates being 1975. Estimates of expenditure on gross national product at constant (1975) prices are recorded in the annual national income book and reproduced here, slightly modified, as Table 6.6.

Table 6.6   Expenditure on gross national product at constant (1975) market prices (£m)

Category	1975	1976	1977	1978
1. Personal expenditure on consumers' goods and services	2,369.9	2,413.7	2,510.8	2,731
2. Net current expenditure by public authorities	688.1	728.3	744.9	777
Total current domestic expenditure	3,058.0	3,142.0	3,255.7	3,508
3. Gross domestic fixed capital formation	870.3	927.6	986.0	1,157
4. Value of physical changes in stock	−53.3	9.8	+ 88.4	+ 19
5. Exports*	1,619.0	1,749.6	1,986.1	2,239
6. *Less* Imports*	−1,814.0	−2,074.6	−2,344.0	−2,701
Gross domestic product at constant market prices	3,680.0	3,754.2	3,972.2	4,222
Net factor income from the rest of the world	21.0	−1.7	−15.4	−17
Gross national product at constant market prices	3,701.0	3,752.5	3,956.8	4,205
Index of GNP at constant market prices (1975 = 100)	100	101.4	106.9	113.6

*Excluding factor income flows.
Source: *National Income and Expenditure 1978*

A comparison of Tables 6.5 and 6.6 will show the effects of price changes on the value of GNP and the principal categories of expenditure. The percentage growth of GNP in real terms is shown by the last line of the table, which relates GNP in each year to the value of GNP in the base year 1975.

Different methods are used to revalue current expenditure categories at constant prices. In the case of personal expenditure, individual constituents are directly revalued at 1975 prices or in some cases deflated by appropriate price index numbers. A breakdown of personal expenditure at constant prices is recorded in the accounts. Other expenditure categories in Table 6.6 are revalued by deflating current expenditure figures by appropriate price index numbers.[18]

An alternative approach, also used by the CSO, is to start from the output side by estimating the contribution of each sector (agriculture, industry, etc.) to gross national product, valued at constant prices, and then adjusting to market prices by adding on taxes on expenditure and subtracting subsidies, also valued at constant prices. This approach is similar to that involved in the calculation of value added or net output at constant prices, discussed in Chapter 3 in relation to agriculture and Chapter 4 in relation to industry. One method of measuring value added at constant prices (constant factor cost) is by 'double deflation'. This involves re-valuing the output of the sector at constant (base-year) prices, and then re-valuing the inputs (of materials and services) of the sector at constant prices. The difference between output and input at constant prices is value added at constant prices. Alternatively, an index of the volume of production for the sector may be compiled; the index for any year is then applied to the base year figure of value added to give value added at constant prices for the year in question.[19] The former method is used in Ireland to estimate the contribution of agriculture to gross national product at constant prices, while the latter is widely used for industry and some other sub-sectors of the economy.

For other sectors, notably services, public administration and defence, neither method is easily applicable and alternative methods have to be used, often based on employment. In public administration, for instance, in which value added comprises the wages and salaries of civil servants, an index of the volume of production can be compiled on the basis of changes in the number of civil servants employed, with or without any adjustment for productivity changes. Such an index can then be applied to the base year wage and salary bill to yield an estimate of value added at constant prices.

Estimates of GNP at constant prices by sector of origin, and comparisons of expenditure-based and output-based constant price GNP estimates, are recorded in the national income book. Differences between these are small, though undoubtedly revisions are made in the course of preparing the estimates to effect the maximum degree of reconciliation.

Figures of GNP at constant prices are more appropriate for analysing year-to-year changes in the standard of living, since price changes have been eliminated and the figures refer to changes in the volume of goods and services produced. Of most immediate concern to the standard of living are changes in the volume of goods and services available for personal and collective consumption. In Table 6.6, items 1 and 2 have been added together to give total supplies of current goods and services for domestic use. Subject to qualifications discussed below, this aggregate provides a guide to changes in the standard of living.

An important factor in assessing changes in the standard of living is the effect of changes in the size and structure of the population. If population is increasing at the same rate as real national product then real GNP per capita will remain constant. Consequently, changes in real GNP should be examined in conjunction with changes in population. For example, although real GNP in Ireland in 1958 was no higher than in 1953, the decline in population over the period resulted in a small increase in real GNP per capita. In countries like India, on the other hand, with a rapidly increasing population, increases in output have been barely sufficient to maintain the already low level of GNP per capita.

Also relevant to analysing changes in real GNP are more complex factors such as the structure of the population (the average size of family and the proportion of gainfully-occupied persons in the population) and the distribution of income. Changes in the distribution of income which make some people relatively better-off and others relatively worse-off affect the economic welfare of the community in a way which cannot be shown in statistics of gross national product or personal expenditure.

Statistics of national income and GNP (at current and at constant prices) per head of population are published in the annual national income book. Comparable figures per person at work are also published. Changes in GNP per head of population are of interest in analysing changes in the standard of living. Changes in GNP per person at work are useful in indicating general trends in productivity, and for comparisons with other countries.

National income and social accounting

The framework of the national accounting system as described above rests upon the identity of income, output and expenditure. Within that framework the accounts are concerned with three forms of activity — production, appropriation and accumulation — and four categories of transactors — enterprises, government, persons and the rest of the world. Moreover the accounts are designed to measure real flows as well as financial flows between these transactors.

The relations between different transactors and forms of activity may be shown in greater detail by means of a set of *social accounts*.[20] An account is drawn up for each class of transactor showing on one side of the account the sources of income (credit items), and on the other the distribution of that income (debit items). A separate account may be prepared for each class of transactor and form of activity. In the case of enterprises, for example, the production account will show the sources of income (receipts from sales) as credit items, whilst the debit side of the account will show the principal categories of expenditure (wages and salaries, imports of materials, gross profits: since all enterprises are grouped together the purchase and sale of materials between enterprises is netted out). The appropriation account of enterprises includes as a credit item the gross profits of enterprises, and on the debit side the distribution of these profits between taxes, dividends and company savings. The latter is then carried forward to the capital account of enterprises, showing on the credit side the sources of investment funds, and on the debit side the distribution of these funds as investment expenditure. Similar accounts may be drawn up for persons, government and the rest of the world.

In principle, the two sides of any account should be equal, bearing in mind that borrowing and lending are included as credit and debit items respectively. Each item in the set of accounts should appear twice, once as a credit item and once as a debit item. For example, the dividends paid by enterprises will appear as a debit item in the appropriation account of enterprises, and also as a credit item in the appropriation account of persons. Again, the savings of enterprises will appear as a debit item in the appropriation account and be included as a credit item in the capital account of enterprises.

These two features of the social accounts explain their usefulness. They are designed to show the relationships between different sectors and forms of activity, and to ensure consistency in the treatment of transactions. For example, if the value of changes in livestock on farms is included as part of personal income, it must also be included as a debit item in the capital account of persons. If the constituent items of the accounts are independently estimated, the preparation of the accounts is a useful check on the consistency and accuracy of the estimates, since the structure of the accounts requires consistency in the treatment of particular items and for the identification of flows between sectors and forms of activity. If estimates of items of income and expenditure are independent, there are almost certain to be some discrepancies which will show up in the accounts in the form of inequalities between credit and debit items. It will then be necessary to make adjustments to some of the items in order to balance the accounts.

This system of social accounts can play an important part in the presentation and construction of the national accounts. The main tables of income and expenditure are built up from the social accounts through the consolidation of individual social accounts and the 'netting out' of identical items on both sides of the accounts. Some of the tables in the national income accounts are presented in the form of social accounts, in particular the tables of personal income and expenditure, income and expenditure of public authorities, and savings and capital formation. As an example, Table 6.7 shows the appropriation account of public authorities.

Table 6.7 Appropriation account of public authorities, 1978 (£m)

Credits		*Debits*	
Taxes on income and wealth (incl. social insurance)	1,014	Subsidies	242
Taxes on expenditure (incl. rates)	1,045	Transfer payments and national debt interest	1,170
Net trading and investment income	171	Net current expenditure on goods and services	1,130
Current transfers from the rest of the world	12	Public authorities' savings	—300
Total	2,242		2,242

Source: *National Income and Expenditure, 1978*

The left-hand side of the table shows the sources of income of public authorities, and the right-hand side shows the disposition of that income. Note that transfers between public authorities — for instance, central government grants to local authorities — are omitted in the consolidated set of accounts. Public authorities' expenditure exceeded income in 1978, and the difference of £300m is shown as a negative debit item, funded by public authorities' borrowing. This £300m will also be shown in the accumulation account of public authorities as a negative credit item.

A particular form of presentation of the national accounts, which illustrates the relations between income, output and expenditure and emphasises the flow of goods and services between different sectors of the economy, is the inter-industry or input-output table. The basic inter-industry relations (or input-output) table is less comprehensive than the system of social accounts, but much more detailed within its own sphere. It is concerned with only one form of economic activity, production, and one category of transactor, enterprises. It is in fact a detailed breakdown of national product by sector of origin, in which inter-sector flows have not been netted out, and is designed to show directly the relation between the production of goods and services and the allocation of these goods and services between different categories of expenditure.

A simple hypothetical input-output table is shown in Table 6.8. The different sectors identified in the table are those which appear in Table 6.2, showing net national product by sector of origin, but the figures in the table are imaginary. Transactions with the rest of the world are excluded — it is assumed that there are no exports or imports.

The gross output of a sector or industry consists of the value of materials bought from other industries plus the value added by the industry itself, as shown in the first five columns of the table. Sales and purchases between establishments in the same industry have been netted out. The output of each industry is sold either to other industries (as 'inputs') or to final buyers: these demands by final buyers constitute the familiar categories of personal expenditure, government current expenditure and capital formation. Since for each industry receipts = expenditure, the totals for each corresponding row and column must be equal.

Examination of the table will reveal the identity between national income, national product by sector of origin and expenditure on national product and the relation, in terms of flows of goods and services, between the income and expenditure sides of the accounts. The second last row of the table shows the value added by each industry, and hence the contribution of each industry to national product; ignoring depreciation, these figures are equivalent to those for net national product by sector of origin, recorded in the national income accounts. The sum of value added by each industry gives net national product at factor cost = national income, which in the example of Table 6.8 is 98. The expenditure or outlay by final buyers is also 98 (60 + 24 + 14). This identity between expenditure and national income by sector of origin follows necessarily from the construction of the table. National product is simply the sum of gross outputs less inter-industry sales. Since inter-industry sales = inter-industry purchases, national expenditure = national product.

Table 6.8 Inter-industry relations

From \ To	Intermediate sales and purchases						Sales to final buyers			
	Agriculture	Industry	Distribution and transport	Other Domestic	Public Administration and Defence	Total intermediate	Personal consumption	Govt. expenditure	Capital expenditure¹	Gross output
Agriculture	—	20	—	5	—	25	21	—	4	50
Industry	15	—	12	10	2	39	14	2	5	60
Distribution and transport	5	10	—	4	3	22	6	1	1	30
Other Domestic	2	3	2	—	4	11	17	4	3	35
Public Administration and defence	—	—	—	—	—	—	2	17	1	20
Total intermediate	22	33	14	19	9	97	60	24	14	
Value added²	28	27	16	16	11					
Gross output	50	60	30	35	20					195

¹Capital expenditure includes stock changes.
²Includes wages and salaries, profits and all other factor payments.
Since sales and purchases between establishments in the same industry have been netted out, the main diagonal of the intermediate part of the table contains zero entries. In addition there are other zero entries where it is assumed that no transactions take place between the industries concerned. The sector 'public administration and defence', for instance, is assumed to sell all its 'output' to final buyers.

The merits of the inter-industry table in aiding the formulation of national accounts lie in its detailed analysis of the flows of goods and services between sectors, and the requirement that for each industry incomings (sales) must equal outgoings (purchases). Given the data on gross outputs and intermediate purchases, the columns of the table may be built up, and expenditure by final buyers entered as residuals to equate the row and column totals for each industry. These residual estimates may then be compared with direct estimates of expenditure on national product, and adjustments made in the event of discrepancies between the two sets of estimates. In addition, estimates of value added may be compared with independent estimates of national income. There are many ways in which tables of inter-industry relations may be used within the framework of the national accounts.

In practice, the construction of input-output tables is complicated by the need to include imports, indirect taxes and subsidies, and the table itself usually consists of a large number of industries for which a considerable amount of data is required, obtained from sources such as the Census of Industrial Production, the annual trade returns and similar sources. These sources are by no means comprehensive and it is difficult in many cases to obtain accurate information on inter-industry sales and purchases. Input-output tables, however, are very useful not only as a supplement to the national accounts but also in analysis of the structure of the economy, projections of the future economic structure, and various other purposes.

Input-output tables for Ireland for the years 1956, 1964 and 1969 were compiled by the CSO, and for the years 1968, 1974 and 1976 by E. W. Henry.[21] A good introduction to input-output analysis and its applications, including applications with Irish data, is O'Connor and Henry (1975).

Reliability of estimates

The estimates published in the national income and expenditure accounts are based on a variety of sources, almost all of which are subject to error. As the CSO states, 'more reliance can be placed on the changes between consecutive years than on the absolute level of any single figure'. The text of the national income book is not forthcoming about the precise methods of estimation, and it is difficult to assess the accuracy of the statistics. Methods of estimation for some of the items are included in the explanatory notes to the tables, and other information is sometimes to be found in the text, particularly when there have been changes in the methods of estimation of certain items. For example, in 1972 there was a change in the method of treating emigrants' remittances, and in 1976 there was a major change in the treatment of banks and other financial institutions. These changes are described in the national income and expenditure reports for those years.

In theory, a check on the accuracy of the statistics is provided by the necessary identity between national income, net national product and expenditure on national product. In practice, this check is ineffective because the estimates are not independent. Two of the most important constituent items, personal expenditure and personal savings, are residual estimates which are fixed with the precise intention of establishing the identity between income, output and expenditure. A result of this method of estimation is that personal expenditure will reflect any errors in the estimates for other items, in particular the estimates of personal income. The CSO does in fact make direct estimates of personal expenditure, which are then adjusted to agree with the required residual total. It would be of interest to know the difference between this direct estimate and the residual total although, in view of the difficulty of estimating personal expenditure directly, it is quite likely that the residual estimate is more accurate.

Estimates for other expenditure categories are subject to smaller margins of error. Fairly accurate statistics of imports and exports are built up from the trade returns and other sources (see Chapter 5). Estimates of public authorities' expenditure are derived from the detailed financial accounts of central and local government. Estimates of gross capital formation, which are more difficult, are built up on the basis of output statistics of capital goods industries, imports of capital goods, and statistics of building and construction. Changes in the value of stocks and work in progress, also based mainly on census of production data, and the corresponding adjustment for stock appreciation are difficult to estimate accurately. However, they account for only a small proportion of total outlay. Whilst the margin of error in this estimate may be quite large, the absolute magnitude of any error is unlikely to be significant. The difference between these items of expenditure and the 'control total' of gross national product at market prices is personal expenditure on current goods and services.

The income estimates form the basis of the national income and expenditure accounts. Estimates of the constituent items of national income — as a glance at Table 6.1 will suggest — will clearly be of varying reliability. Estimates of company profits are based on analysis of a sample of company accounts weighted towards larger companies. With such a coverage, it is reasonable to expect that estimates of company profits are good.

Agricultural income is calculated as the difference between estimated gross receipts from sales of agricultural produce and estimated expenses such as materials bought, rates, rent etc. Estimates are also made of the value of produce consumed on farms, payments in kind and the value of changes in livestock. These estimates are based on a variety of sources of agricultural statistics, such as the census of agriculture.

Estimates for some of the other items in Table 6.1, such as social insurance contributions and the income of the Post Office, are fairly straightforward, and sources of estimation for net income from abroad are referred to in Chapter 5. The most important of the remaining items are 'Other trading profits, professional earnings, etc.' (item 4 in Table 6.1) and 'Wages, salaries and pensions' (item 5). Estimates for these items are based on data provided by the Revenue Commissioners from tax returns, the censuses of production, agriculture, distribution and population, and the regularly published statistics on employment and earnings. In view of the complexity of estimates of this nature, the margin of error in the calculations, though perhaps not large in percentage terms, may be quite large in absolute terms. It should not be concluded from this that national income statistics are 'unreliable'; margins of error are inevitable in estimates of this kind, however good the methods of estimation. In any case, reiterating the point raised at the beginning of this section, for most purposes the trend in the figures is more significant than the absolute level of any estimate.

Other sources

The main tables in the national income accounts include estimates for the current year and for the preceding six to seven years. Estimates for the current year, rounded to the nearest million pounds, are provisional, and there are invariably some adjustments to the figures which are reflected in the following year's publication. In addition, there are sometimes changes or revisions in the methods or concepts involved in the estimates for particular items which, to ensure consistency in the series, necessitate revisions in the estimates for earlier years. Moreover, as a result of the interdependent structure of the accounts, revisions in one item require corresponding adjustments to other items, particularly residual items like personal expenditure and personal savings. For an analysis of revisions to the accounts, their effects and implications, see Ruane (1975).

Official estimates of national income and expenditure date from 1938, but some private and semi-official estimates were made for earlier years. The first estimates were made by T. J. Kiernan, for the year 1926[22]; independent estimates for 1926 and for the years 1929, 1931-40 were made by Professor G. A. Duncan.[23] and alternative estimates for 1929 and 1933 were made by the statistics branch of the Department of Industry and Commerce.[24] These estimates represented some of the earliest attempts to define and measure national income and expenditure, and Ireland has been one of the pioneering countries in this field.

The annual national income book is the principal source of statistics on national income and expenditure. Other sources include the annual pre-budget statement issued by the Department of Finance and the *Review and Outlook* published by the Department later in the year, and the publications of the Central Bank. The Central Bank's *Annual Report* includes provisional estimates of national income and expenditure for the year just ended, as well as a commentary on trends in the major aggregates. This is a useful source of information on trends in income and expenditure, since it may be read in conjunction with the Bank's analysis of recent financial and monetary developments. The *Quarterly Bulletin* of the Bank is also a useful source for analysis of short-term trends, as is the *Quarterly Economic Commentary* published by the Economic and Social Research Institute. A recent innovation by the Research Department of the Central Bank is the estimation of the major national accounts aggregates (particularly expenditure categories) on a quarterly basis. Quarterly estimates are valuable for the evaluation of current trends and for short-term forecasting, particularly for modelling the behaviour of financial and stock variables in which adjustment lags are typically shorter than one year. For a discussion of the methodology and estimates of quarterly national accounts aggregates 1963-1977, see O'Reilly (1981).

Notes to Chapter 6

[1]   Prl. No. 7356 (Stationery Office 1945). Private estimates of national income and expenditure were made for some earlier years. See section on other sources below.

[2]   *Tables of National Income and Expenditure, 1938 and 1944-1950* Pr. No. 350.

[3]   Suppose a company borrows £100,000 from a bank on which 15 per cent interest is due. According to the standard convention the trading profit of the company will be defined to include the £15,000 interest due to the bank. The trading profit of the bank will, however, be defined to include that £15,000 interest *less* whatever interest the bank paid its depositors — say 10% or £10,000. Thus there will be double-counting of £5,000.

[4]   Following the example of footnote 3, the adjustment for financial services would be − £5,000.

[5]   So-called because, through the tax system, they effect a redistribution of factor incomes.

[6]   By convention, it is assumed that no tangible assets correspond to the national debt, so that interest payments on government stock cannot be regarded as a form of return on capital employed, or as part of factor incomes. Interest payments are financed by general taxation, and are thus treated as transfer payments. To the extent that at least in part the national debt reflects the existence of tangible public-owned assets (such as roads and buildings) this assumption is not strictly valid, but it would clearly be difficult to estimate with any accuracy the value of imputed income of this kind.

[7]   This consists of a miscellany of items including the trading income of the Post Office, income from government investments in industry, the net rental income of local authorities, and other similar items.

[8]   Except for incomes arising in agriculture, which are based on output estimates. Thus the estimates of income by sector or origin published in the national income book are not independent estimates, but a re-classification of the income estimates shown in Table 6.1.

[9]   In underdeveloped countries with a poor infrastructure, a limited fiscal system and an extensive use of barter and income in kind, output estimates may be the only feasible way of measuring national income. The 'direct' method of estimating actual factor incomes presupposes a fairly efficient and extensive fiscal system, and in general a reasonably advanced level of economic development.

[10]   This identity is fundamental to the structure of the national accounts. Since the principal concern here is with sources and methods discussion of the theoretical aspects of national income accounting has been kept to a minimum; understanding of the structure of the national accounts, however, is facilitated by some knowledge of the elementary theory of income determination. See for example Samuelson (1976).

[11]   In the UK income and expenditure accounts the difference between the two sets of accounts is never wholly reconciled, and the difference is entered as a "residual error" on the income side. This, however, should not be taken to mean that the two sets of estimates are wholly independent — nor should it be taken as a true measure of the actual difference between the two sets of estimates, since prior adjustments will already have been made.

[12]   Purchase of cars for business use are however included in gross fixed capital formation.

[13]   Apart from their inherent benefits, a better-educated and healthier community is likely to be more productive so that 'social' investments may also be regarded as materially productive in the long-term.

[14]   However information on investment expenditure is collected in the census of industrial production (see Chapter 4) and in the estimates of agricultural output (see Chapter 3). For other sources and studies, see Kennedy and Dowling (1976) and Vaughan (1980).

[15]   In an open economy this relationship is slightly more complicated; the equality between savings and investment is maintained by net foreign borrowing or lending. Thus investment = domestic savings + net borrowing from abroad (or net lending abroad, which is negative). See following text.

[16] Let Y = income, C = consumption, S = savings, I = investment, E = expenditure. Then Y = C + S, E = C + I, and since Y = E, C + S = C + I, hence S = I.

[17] Or by running down Irish-owned assets abroad — hence the term 'net foreign disinvestment'. In practice however the willingness of foreigners or foreign companies to invest or lend in Ireland has been sufficient to offset any deficit on current account. For a more detailed discussion of this see Chapter 5.

[18] References to the methods used, as well as tables are included in the text of the national income book. Also included are index numbers of each expenditure category at constant prices, and an adjustment ot GNP at constant prices to allow for changes in the terms of trade. The latter is designed to account for the effects of relative changes in export and import prices on real domestic purchasing power, and hence upon the real volume of GNP. See Chapter 8.

[19] Thus suppose value added in the cement industry in 1975 was £10m and the index of the volume of production of the cement industry in 1978 was 120. Value added in 1978 at constant (1975) prices can be calculated as £(10 × 1.20) = £12m.

[20] For an exposition of social accounting systems see United Nations (1968).

[21] CSO (1970), Input/Output Tables for 1964 (Prl. 985); CSO (1978), Input/Output Tables for 1969 (Prl. 5383); E. W. Henry (1972), Irish Input/Output Structures 1964 and 1968, *ESRI Paper No. 76;* E. W. Henry (1980), Irish Input/Output Structure 1976, *ESRI Paper No. 99.*

[22] *Economic Journal* March 1933 and *JSSISI* 1933. The former includes estimates of national income, and the latter estimates of national expenditure.

[23] 'Social Income of the Irish Free State 1926-38', JSSISI, Vol. XVI, 1939-40. 'Social Income of Eire 1938-40', JSSISI, Vol XVI, 1940-41. Earlier estimates, by the same author, for the years 1929 and 1931-35, are included in the *Report of the Commission on Banking, Currency and Credit* (1938) – Chapter 2 and Appendix 7.

[24] Commission on Banking – Appendix 8.

7 Income, Wealth, Expenditure and Taxation

The last chapter focussed on the problems inherent in estimating national income or output. This chapter starts with the estimates of personal income (see Table 6.2 above) presented in the national accounts, and describes the estimation of per capita personal incomes on a county and regional basis using the accounting conventions set out in Chapter 6. Estimates of household incomes derived from the *Household Budget Survey* will be reviewed, and the differences between this concept and that of personal income explored. Income estimates are of limited use in the absence of data on income distribution or inequality. Studies of the distribution of personal income in Ireland are reviewed. The progression from income is to the broader concept of wealth where studies of the ownership and distribution of personal wealth in Ireland are described. Income and wealth both generate expenditure and the basic data sources, the Household Budget Survey and the Retail Sales Index, are reviewed in some detail. Studies of income and expenditure relations in Ireland are also described. The chapter concludes with an overview of the available data on taxation.

Income

Two distributions of income can be defined: personal income distribution and functional distribution of income. The latter denotes the distribution of income between the various factors of production. This section focusses on personal income. Statistics of income in Ireland are not well developed, with the exception of the National Accounts. Available estimates of county and regional incomes and of income distribution have been compiled by private researchers. Their main aim has been to estimate per capita personal income by county by distributing the National Accounts aggregate of personal income between counties or planning regions. Initiated by Attwood and Geary

(1963), this work has been extended by Ross (1969, 1972, 1980) and by Ross and Jones (1977). The underlying methodology is detailed in Ross (1971).

Personal income can be regarded as an approximation to the purchasing power of an area. It includes not only income generated within the area but also payments from outside for factor services used elsewhere but belonging to residents of the area, plus current transfers from the central government. The personal income of a county will therefore include the income both of households and of private non-profit-making institutions. The concept of personal income is thus much broader than simply wages and salaries. It includes not only the remuneration of the employed and self-employed (including in the case of farmers the value of changes in livestock numbers) but also income received from interest, dividends and rents (both actual and imputed), current transfers from public authorities and emigrants' remittances and pensions.

The methodology employed assigns each component of the National Accounts aggregate 'Personal Income' to the county in which its recipient resides. Control totals detailing employment by county from the 1971 *Census of Population* and the *Census of Industrial Production* were used in the compilation of the 1973 and 1977 estimates. Since the former source refers to April and the latter to September, problems arise in some industries with respect to seasonal employment and to workers holding more than one job. Commuters also pose problems, particularly for those counties containing large urban centres. Large numbers of workers may be employed in a county other than that in which they reside.

While a satisfactory methodology has been devised to allocate employee income to the county of residence, no satisfactory solution exists for the distribution of the National Accounts aggregate of 'income from self-employment' across individual counties. In the existing estimates such income has simply been distributed on a per capita basis in accordance with the number of self-employed returned in each county by the Census of Population. Further progress in this area may be possible with the computerisation of the Revenue Commissioners' records of those paying tax under Schedule E. However, the quality of the resulting data will be diluted somewhat by the Revenue practice of assigning taxpayers not to their county of residence, but to the tax district in which they reside. The boundaries of tax districts and those of counties regrettably do not coincide.

Estimates of personal income at county level now exist for 1960, 1965, 1969, 1973 and 1977. (The 1969 estimates are not, however, directly comparable with those of other years, see Ross and Jones (1977)). The estimates compiled by Ross et al relate to gross income before the deduction of direct taxes and employees' social insurance contributions.

Estimates of personal income are not a perfect measure of economic welfare. They inevitably reflect the breadth of the market sector in different counties and the varying incidence of direct taxation. The latter may be of particular importance when assessing changes in inter-county income differentials over time, given the initial low rate of direct taxation on the agricultural as compared to the non-agricultural sector during the 1960s and 1970s. The definition of personal income embraces the income of private non-profit-making bodies. In the 1973 estimates the household element of personal income has been isolated by deducting transfers to these bodies, yielding a somewhat more satisfactory measure of economic welfare. Even this refined concept has drawbacks because it takes no account of the impact of direct taxation. At present no satisfactory estimates of household *disposable* income can be derived by this method.

The only other comprehensive source of data on incomes is the country-wide *Household Budget Survey* of 1973, whose main purpose was to determine in detail the current pattern of household expenditure in order to provide a comprehensive weighting basis for the consumer price index. The most significant feature of the 1973 survey was the inclusion, for the first time in an Irish household budget survey, of rural as well as urban households. Previous inquiries, 1951/1952 and 1965/1966, were confined to urban areas only. The 1973 survey results are of particular interest because they give for the first time expenditure and income data based on a sample of all households within the state.

The survey of 1973 embraced a final sample of 7,748 households. Within each household all members aged fifteen years and over were asked to maintain detailed records of their day-to-day expenditure for two consecutive weeks and to provide detailed information about their income and regular personal expenses. In addition, sample households located in rural areas and in towns of fewer than 1,000 inhabitants who were operating medium to large farms were asked to maintain detailed farm accounts over a twelve-month period.

The concept of income employed is that of 'household income'. This differs in a number of important respects from the national accounts concept of personal income, and accordingly from the definition of personal income employed by Ross et al. The Household Budget estimates relate to the household, while Ross's relate to the individual. Furthermore, Ross's figures are consistent with and derived from the national accounts, but the HBS estimates are not. In particular, they exclude the imputed rent of dwellings and the value of changes in livestock numbers.[1] The two sets of income data are therefore not compatible.

The published results of the HBS, 1973, distinguish household income by region. The degree of detailed disaggregation of income is considerable. Weekly household income is classified by household size, by the socio-economic group of the head of the household and by household tenure. The following components of weekly household income are then distinguished: direct income,[2] state transfers and direct taxation. The sum of all three yields disposable income. Direct income is further sub-divided into earned income (separately distinguishing employee income and self-employed income from both non-farm and farming enterprises), retirement pensions, investment income, property income, the value of own garden or own farm produce at retail prices, and other direct income. State transfer payments are sub-divided into childrens' allowances, old age and retirement pensions, widows and orphans' pensions, unemployment benefits and assistance, education grants and scholarships, and other state transfers. Direct taxation is sub-divided into income tax and social insurance deductions.

A similar disaggregation of household income is presented in the continuing annual Household Budget Surveys introduced in 1974. However, the continuing inquiry relates only to urban households. Four categories of towns are distinguished: Dublin and Dun Laoghaire, other towns with more than 10,000 inhabitants, towns with 5,000 to 10,000 inhabitants and towns with 1,000 to 5,000 inhabitants. In interpreting income data from such sources, note that respondents typically tend to underestimate household income. There is an understandable reluctance to disclose full details of household income to an unknown interviewer. The results of the *Family Expenditure Survey* in the UK are similarly characterised. In Ireland, the degree of understatement is estimated to have reached 15 per cent on occasion.

Other less comprehensive sources of data on incomes include surveys of farm incomes carried out since 1955 — initially as the *National Farm Survey* by the CSO, and in more recent years as the *Farm Management*

Survey conducted by An Foras Talúntais[3] (see also Chapter 3 above). The principal concept of income employed in the Farm Management Surveys is that of management and investment income, i.e. the return to the capital and entrepreneurial investment made by the farmer. It consists of the residue of income after the labour input, both hired and family, has been compensated at standard agricultural wages. The Farm Management Survey presents estimates of this income concept by farm soil type and by four categories of farming system: mainly dairying; dairying and dry stock; dairying, dry stock and tillage; mainly dry stock. The survey also records the size of each farm in acres and the size in acres adjusted for any uncultivable land. It is thus possible to calculate management and investment income per adjusted acre for farms in each county by soil type. Similar calculations show income per labour unit and per family labour unit.

Income Distribution

The availability of information on income distribution in Ireland is limited, and that which is accessible is restricted both in scope and definition. Data on the distribution of personal income is particularly lacking. It is consequently almost impossible to assess the existing degree of inequality in the distribution of personal income in Ireland, or to evaluate the distributional consequences of government taxation and spending programmes.

The first attempt to estimate the distribution of personal income in the Republic of Ireland was made by Reason (1961). His estimates were based on income tax data, employers' tax returns, Census of Industrial Production returns and payments made by government departments to their employees. His estimates relate to the distribution of non-agricultural incomes in 1954. Geary (1977) and Norton (1976) have used the results of the 1965-6 Household Budget Enquiry to derive estimates of the distribution of average weekly household income in urban areas. Norton (1976) used the results of the Farm Management Survey of 1966 – 67 to provide an estimate of the distribution of family farm income. The National Economic and Social Council (NESC; 1975D) has reviewed the available evidence on income distribution from the complete range of Household Budget Surveys up to 1973, and explored the possibility of using alternative data sources, e.g. the Revenue Commissioners' Database, the Farm Management Survey, and earnings data generated by the Census of Industrial Production, to remedy some of the deficiencies identified.

The Household Budget Survey of 1973 has been exploited by Nolan (1978A) to estimate the distribution of both direct and disposable personal income in Ireland. He also used it to contrast the patterns of distribution in urban and rural areas, to assess the overall impact of state transfers and taxes on income distribution and to compare the distribution in the different planning regions. The trend in urban distribution from 1965-6 to 1976 is also examined.

Nolan concluded that the distribution of income in rural areas was considerably more unequal than in urban areas. State transfers and direct taxation were found to lead to a more equal distribution, with state transfers having the greater effect. The distribution of disposable income in urban areas appeared to be more equal in 1976 than a decade earlier.

The redistributive effects of taxes and transfer payments have also been analysed by the CSO, using the same data source.[4] This study concluded that the overall effect of the state's intervention was to redistribute income from high income to low income households. This broad conclusion is sensitive to household composition. The income level up to which benefits exceed taxes paid increases progressively as the number of adults and children in the household rises. This reflects correspondingly greater imputed values for health and educational services.

Many studies of the distribution of income or wealth utilise a graphical device known as the *Lorenz curve,* and a summary statistic based on it, the *Gini coefficient.*[5] A Lorenz curve is drawn by ranking income recipients from the poorest upwards on a cumulative basis. The axes on which the data are plotted are graduated from 0 to 100, the horizontal axis measuring the percentage of gross (i.e. total) income recipients, the vertical the percentage of total income. The measurement on both axes is cumulative. The Lorenz curve therefore plots the percentage of income recipients against the percentage of total income which accrues to those recipients.

This process can best be illustrated by means of Table 7.1, drawn from Nolan's (1978A) study. The first column records the percentage of households in each income group. The second records the percentage of gross income accruing to households in each group. Thus, 10.467 per cent of all households sampled in the 1973 survey had an income in the range of £10 to £15. The combined incomes of all households in this range accounted for 3.496% of gross income in the state. Columns 3 and 4 of the table simply cumulate these percentages. Thus taking the first two lines of the table, that is the 8.266% of households earning less than £7 per week whose combined incomes accounted for 1.072% of gross income, and the 5.739% of households with incomes in the range of £7 to £10 whose combined incomes amounted to 1.319% of gross income, it can be seen that the poorest 14.005% of households received only 2.391% of total gross income.

Line 6 reveals that 52.483% of households cumulated from the poorest upwards accounted for only 23.347% of gross income. The very last line reveals that the 7.8% of households whose weekly income at the time of the survey was in the range of £80 and over received almost a quarter of total gross income. The graphical representation of the two cumulative percentage columns yields the Lorenz curve of Figure 7.1.

Table 7.1 Distribution of gross income 1973

Gross income range per week	Percentage of households	Percentage of income	Cumulative percentage of households	Cumulative percentage of income
Under £7	8.266	1.072	8.266	1.072
£7 — under £10	5.739	1.319	14.005	2.391
£10 — under £15	10.467	3.496	24.472	5.887
£15 — under £20	8.326	3.960	32.798	9.847
£20 — under £25	9.040	5.529	41.838	15.376
£25 — under £30	10.645	7.971	52.483	23.347
£30 — under £40	14.065	13.270	66.548	36.617
£40 — under £50	9.753	11.817	76.301	48.434
£50 — under £60	7.404	11.048	83.705	59.482
£60 — under £70	4.728	8.339	88.433	67.821
£70 — under £80	3.747	7.593	92.180	75.414
£80 and over	7.820	24.586	100.000	100.000
Total	100.000	100.000		

Source: Adapted from Nolan (1978) Table 7. Original data from *Household Budget Survey* 1973, Volume 4, Table 7.

Perfect equality in the distribution of income would obviously give a Lorenz Curve coinciding with a straight line of 45 degrees, i.e. the line of perfect equality. The extent to which income is unevenly distributed

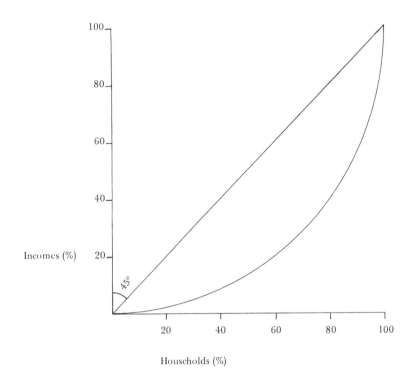

Figure 7.1   Distribution of income; Lorenz diagram of data in Table 7.1

can therefore be gauged by measuring the area between the Lorenz curve and such a line of perfect equality. The more unequal the distribution of income the larger will be the area between the line of perfect equality and the Lorenz curve. The Gini coefficient is defined by the ratio of the area between the Lorenz curve and the 45° line to the total area under the 45° line. The coefficient ranges from zero, where there is perfect equality in the distribution of income, to unity, where all income accrues to one household. The Lorenz curve may also be employed for comparative purposes by superimposing curves plotted for different points in time or for different geographical areas. In such circumstances, however, the calculated Gini coefficients may be of only limited use for comparative purposes. Where Lorenz curves intersect, the ranking of the underlying distributions by their Gini coefficients does not provide an unambiguous verdict on their relative degree of equality.[6] Such a conclusion can only be drawn where one Lorenz curve lies wholly above or wholly below the other (see Atkinson (1974) and Paglin 1975)).

Wealth

Inadequate though the data on income distribution is, less is known about the distribution of personal wealth. Wealth is a stock, income a flow. Wealth generates income. To a certain extent the distribution of income will mirror the distribution of wealth, but changes in the composition of total wealth will obviously affect the distribution of personal income. Ireland is not unique in the poor degree of documentation of wealth distribution; few countries possess adequate statistics.

A technique has therefore been developed to provide indirect estimates of wealth distribution. Data on the estates of people who die during a particular period are used to estimate the underlying distribution of wealth during that period. On the assumption that those who die are representative of the underlying population, the estates which they leave can be grossed up to provide an estimate of the distribution of wealth in the population as a whole. The methodology is relatively simple. The deceased in any year are assumed to be typical of their parent population; given data on the deceased by age, sex, and size of estate it is possible, using *mortality multipliers* derived from age-specific mortality rates, to provide an estimate of the underlying distribution of total wealth. The technique has been applied to Irish data by Lyons (1972) to provide an estimate of the distribution of wealth in 1966. Lyons' estimates are disaggregated by age, sex, and county of residence.

Lyons (1975) examined three major drawbacks of this approach. Firstly, the mortality multipliers used to gross up the estate duty data to estimate total wealth might not be correct. It appears that multipliers based on the mortality experience of the population as a whole tend to underestimate the number of wealthy persons, producing errors in the size distribution of personal wealth. This occurs because of the unavailability of data on the correlation between mortality experience and wealth. It seems generally accepted that mortality experience improves with increasing wealth. However, mortality rates by social status or occupation, which might be used as an approximation to mortality rates directly associated with wealth, are not available in Ireland.

The second weakness identified is that the persons who have died, whose age and sex characteristics are used to gross up the estate duty data, might not be a representative sample of the population within each age and sex group. To the extent that the deceased are unrepresentative, any estimates based on them and their estates will exhibit bias either upwards or downwards.

Finally, to the extent that estate duty is successfully evaded by means of under-declaration of the value of assets in the estate, or by gifts made during the deceased's lifetime with the explicit intention of evasion, the estates declared for probate may not be representative of the wealth possessed by the entire population. Unrepresentative estates would obviously bias the resulting estimates of personal wealth.

Lyons (1975) examined the first of these objections in considerable detail, and concluded that the total amount of personal wealth, and to a lesser extent its distribution, is highly sensitive to the chosen set of mortality multipliers. However, several features of the distribution of wealth in Ireland appear to be relatively robust with respect to the choice of multiplier. The degree of inequality is pronounced over a wide range of plausible multipliers. Lyons estimates that in 1976 the top five per cent of wealth holders owned between 60.3% and 63% of total personal wealth. Males appeared to own by far the largest portion, ranging from 66% of the total to 70.5%. The mean wealth per head of the population was estimated to rise regularly with age, reaching a peak in the fifty-five to sixty-five age group and declining thereafter. Lyons interpreted this to mean that inheritance is an important factor in perpetuating the inequality of wealth distribution in Ireland.

Lyons' work has been subjected to considerable criticism. Harrison and Nolan (1975) noted that the estate duty data used by Lyons covered only 37% of the estates of adults who died in the period considered; and they examined the sensitivity of the original estimates to a modification of Lyons' assumption that the average value of the unexamined estates was £50. They argued that Lyons' conclusion that nearly two-thirds of the adult population owned no wealth in 1966 was in large part due to this assumption. Working from the same set of estate duty data, Harrison and Nolan suggest that the bottom two-thirds of the population, rather than owning no wealth as implied by Lyons, own about 20% of the total. However, Lyons' original finding that the wealthiest 5% of the population own about 70% of total wealth remains intact.

Harrison and Nolan's work has not escaped criticism. Chesher and McMahon (1976), again concentrating on the question of the unexamined estates, argue that both the earlier mentioned estimates of the distribution of wealth are in error. Their results suggest that the poorest two-thirds of the population own approximately 10% of total wealth, while the proportion owned by the wealthiest sections of the community is also considerably less than in Lyons' original estimates. Chesher (1979) has returned to this point. He contrasts and examines the appropriateness of the log normal and Pareto distribution as models of the Irish wealth distributions in 1966, and produces revised estimates of the size of total wealth holdings in Ireland at that date; these are reasonably close to those earlier presented in Chesher and McMahon.

Until recently, estate duty data was the only information source on wealth holdings in Ireland. However, the introduction of a wealth tax in 1975, chargeable at the rate of 1% on the taxable wealth of every assessable person on 5 April of each year, has generated an additional data source. For the years 1975 to 1978, when the tax was in effect, the Annual Report of the Revenue Commissioners provides data on the market value of the wealth of persons above exemption thresholds.[7] The report documents the value of property comprised in the taxable wealth of assessable persons. Property is classified as:

1. property in the state (distinguishing productive property and other property)
2. property outside the state

Assessable persons are sub-divided into individuals, discretionary trusts and private non-trading companies. Certain types of wealth were exempted from the tax, specifically dwelling houses and household effects, livestock, bloodstock, and works of artistic or cultural merit. The net market value of exempted property by type of individual is documented in the report.

Other fragmentary data on aspects of wealth exist. The Household Budget Survey records the ownership of certain consumer durables such as television sets, refrigerators and cars. The Department of the Environment publishes annually the number of vehicles, by horse-power size and type of vehicle, under current licence in each county.[8] The Department of Post and Telegraphs maintains records of the number of telephone lines, both automatic and manual, in each Post and Telegraphs Engineering District[9] and also of the number of television licences current in each county and province.[10] The latter data have been analysed in detail by McCarthy and Ryan (1976).

Expenditure

Data on expenditure can be found in four different sources. The composition of household expenditure is documented in the Household Budget Surveys. Expenditure out of personal income is documented in the *National Accounts,* while expenditure undertaken in outlets mainly engaged in retailing is recorded by the *Retail Sales Index.* The current expenditure of local authorities is documented in the *Returns of Local Taxation.*

The Household Budget Survey[11] embraces the occasional large-scale national survey. The most recent was carried out in 1973 and the continuing small-scale urban Household Budget Survey[12] was initiated in 1974 and carried out on an annual basis thereafter. The most recent (1973) nationwide survey has been described above. The continuing urban Household Budget Survey is conducted on a similar basis, but the sample is considerably smaller. For example, in the 1978 survey, returns were provided by 1,734 households. The sample of households included in the annual survey is chosen from the Labour Force Survey sample of the preceding year.

The results of the surveys document average weekly household expenditure in considerable detail. Ten broad categories of expenditure are distinguished:

Food
Drink and tobacco
Clothing and footwear
Fuel and light
Housing
Household non-durable goods
Household durable goods
Miscellaneous goods
Transport
Services and other expenses

Within each of these categories, there is considerable disaggregation. For example, within the food category over 120 individual items are distinguished. The nationwide 1973 survey provides such detailed expenditure data at the household level by planning region, by household tenure, distinguishing households who are owner-occupiers, those with a mortgage, those renting from a local authority, those renting privately, and those occupying their accommodation rent-free. The results are further disaggregated by six social groups based on the occupation of the household head, and by the livelihood status of the head of the household; distinguishing self-employed, employee, out of work, retired and other.

Additional tables document household expenditure by household size, ranging from one to ten persons; by household composition, distinguishing number of adults and number of children present; by household income, and by a cross-classification of household size and household income. Many of these detailed cross-classifications are also available at the planning region level. The annual survey provides similar data, though of course restricted only to urban households. However, the results are subject to two systematic biases. There is a tendency in all countries to understate household income, and a similar tendency to understate expenditure on alcoholic beverages.[13] In the published results, no adjustments are made for such biases.

The Retail Sales Index[14] records the value and volume of sales in retail outlets. The present monthly series, base $1975 = 100$, distinguishes fourteen categories of retail outlet ranging from grocery shops to department stores. The disaggregated monthly series extends from January 1974, the aggregated series from January 1968. The index is compiled using data supplied by a sample of retailers, drawn from the 1971 Census of Distribution. The respondents detail the value (including VAT) of all goods sold, receipts from services and rentals and the cash value of goods sold during the period under hire purchase arrangements.

The index is designed to measure the level of average weekly retail sales in standard four- or five-week monthly periods. Variations in retail sales due to differences in the calendar composition and the number of trading days in each month are eliminated. The value index is converted to volume terms by deflating by a specially constructed price index derived from the Consumer Price Index, whose relevant commodity prices are reweighted by the commodity sales pattern of retail shops given by the 1971 Census of Distribution. As the Consumer Price Index is compiled only for the mid-month of each quarter, monthly figures for the deflator must be interpolated. Both value and volume indexes are seasonally adjusted using the X-11 method (see Chapter 2).

Trends in the Retail Sales Index provide an indication of the underlying trend in consumers' expenditure. Consumer expenditure incurred at markets, mobile shops and auction sales is not included, nor the sales of newsboys, itinerant traders and self-employed retail agents. So long as such excluded outlets retain a relatively constant share of consumers' expenditure over time, the index will remain a reliable proxy for overall trends in personal consumption.

Retail purchases may be financed by means of a direct cash transaction, by hire purchase or by credit sale. The CSO has conducted an annual inquiry into such hire purchase and credit sale transactions since 1955. Earlier inquiries were carried out for 1938, 1942 and 1943. The inquiry covers finance companies and trading concerns, but the CSO cautions that the list of concerns to which the inquiry is sent cannot be considered exhaustive. Coverage is good in the case of finance companies but somewhat more patchy in respect of trading concerns. The inquiry relates only to those transactions for which a written agreement exists, and therefore does not cover all consumer credit transactions in the Republic.

The results of the inquiry show the number of hire purchase and credit sale agreements, their aggregate retail value and the average retail value of each type of agreement outstanding at the time of the inquiry. Separate details are presented in respect of new agreements concluded during the year to which the inquiry relates. The number of agreements concluded is documented, the amount of deposits paid, the amount of investment or advances secured, the retail value of the goods in question, hire purchase or credit sales charges inherent in the agreement, and the amount of instalments outstanding on foot of the agreement at the end of the year. These data are separately presented for finance companies and trading concerns. New agreements are further disaggregated by the type of good to which they relate (the vast majority being vehicles), the amount advanced in respect of each type of good being separately documented for finance companies and trading concerns. In general, finance company agreements relate to transactions involving motor vehicles, agricultural tractors, contractors' plant and industrial machinery. Such companies also finance the bulk of credit transactions involving the purchase of pedal cycles, agricultural equipment, radio and television sets. The majority of credit transactions involving the purchase of domestic and electrical equipment, cookers, stoves, etc. are conducted through the trading concern dealing in these goods.

A considerable volume of research has been done on Irish expenditure patterns at both macroeconomic and sectoral levels. Since disposable income is either consumed or saved, research into the determinants of aggregate expenditure may analyse the consumption or the savings decision. The latter approach was chosen by Kennedy and Dowling (1970). They examined the causes of changes in the ratio of personal savings to incomes in the period from 1949 to 1968, focussing on the distribution of income between the agricultural and non-agricultural sectors. However, their model fails to explain the significant rise in the personal savings ratio which occurred in Ireland in the middle of the 1970s. This phenomenon has been examined by Kelleher (1977), who augmented the earlier study's identification of the sectoral distribution of income with a variable measuring the ratio of liquid assets to income. The performance of the modified model is considerably superior to that of the earlier study.

Expenditure patterns revealed by the Household Budget Surveys of 1951-52 and 1965-66 have been analysed by Leser (1962, 1964) and Pratschke (1970). Systems of demand equations, using data on the composition of personal consumption from the national accounts, have been estimated by O'Riordan (1975) and McCarthy (1977A). A time series study of the demand for food in Ireland has been conducted by O'Riordan (1976), while trends in expenditure on alcohol in Ireland have been analysed by Kennedy, Walsh and Ebrill (1973); one author also analysed data on the consumption of tobacco (Walsh, 1980B).

The Central Statistics Office provides details of the annual estimated expenditure of visitors to Ireland and of Irish visitors abroad. These are based on the results of the Passenger Card Inquiry — a continuous sample of passengers conducted by the CSO at certain airports, seaports and on cross-border routes. Selected travellers are asked to furnish such details as the reason for their journey, their country of permanent residence, length of stay and estimated expenditure. The responses secured are grossed up, using information on total passenger movements obtained from transport companies and automatic traffic counters at cross-border posts. The CSO warns, however, that in recent years it has become increasingly difficult to ensure that a representative sample of passengers has been chosen. In particular, difficulty has arisen in connection with cross-border movements in private motor vehicles.

The series, published in the *Irish Statistical Bulletin,* documents the expenditure by visitors to Ireland to the nearest £100,000. The expenditure of five categories of overseas visitor is separately distinguished:

air cross-channel
sea cross-channel
continental European
transatlantic
cross-border

Passenger fare receipts by Irish transport companies from persons resident outside the state are separately distinguished. Similar data are provided in respect of Irish residents travelling abroad, the above five categories again being employed. The estimated expenditure of Irish residents travelling abroad is further cross-classified by the reason for the visit, distinguishing tourists, those visiting relatives, those on business trips, and others.

Local authority expenditure data
The primary source of data on the current expenditure of local authorities is the *Returns of Local Taxation,* compiled and published by the Department of the Environment from returns submitted by local authorities. Estimates of local authority expenditure during the current year, and provisional out-turn figures for the previous year, are published by the department in a volume entitled *Local Authority Estimates.* The publication of this compensates for the long delay in the appearance of the Returns of Local Taxation.

Three types of local authority are separately distinguished in both publications, i.e. County Councils, County Borough Corporations and Urban District Councils. The expenditure on each type of authority is disaggregated into eight programme groups, namely:

Programme Group 1 — Housing and building
Programme Group 2 — Road transportation and safety
Programme Group 3 — Water supply and sewerage
Programme Group 4 — Development incentives and controls
Programme Group 5 — Environmental protection
Programme Group 6 — Recreation and amenity
Programme Group 7 — Agriculture, education, health and welfare
Programme Group 8 — Miscellaneous services

There is a number of major discontinuities in the Returns of Local Taxation series, notably:

(1)   In 1976 a new programme classification of expenditure was introduced.
(2)   Since 1974, the period to which the accounts relate has been changed from the financial year to the calendar year.
(3)   There have been changes in the administration and/or funding of particular services, notably those of health and housing, the funding of which has now passed from local to central government.

The Returns of Local Taxation identify the current income of each local authority, separating that intended for capital expenditure from that intended to fund revenue or current expenditure. The capital expenditure of local authorities is usually funded by borrowing from the exchequer through the Local Loans Fund; the annual capital allocation is controlled by central government through the Public Capital Programme. Revenue (or current) expenditure is financed from rates[15], grants from the state and miscellaneous receipts from such sources as housing rents and fees for services. The contribution by the Exchequer to each local authority's rate account, i.e. the agricultural grant, the grant in relief of rates, and the contribution of specific state grants at programme group level for individual local authorities, is distinguished. For local authority-administered services therefore, the absolute and relative level of central government current grants, both general and specific, can be identified.

These data make it possible to analyse and monitor the contribution of central government to regional resource allocation, though the data base is necessarily incomplete, since the contribution of central government via direct expenditure by government departments and agencies is not included in the Returns of Local Taxation. The available data have been analysed by Copeland and Walsh (1976) who investigated inter-county variation in the rate burden and in per capita local authority expenditure.[16]

The Returns of Local Taxation also document the income and expenditure of certain non-rating authorities such as Town Commissioners, Burial Boards, Cemetery Joint Committees, Joint Library Committees, An Chomhairle Leabharlanna and the Loch Corrib Navigation Trustees. The volume is prefaced by a useful introduction to the current system of local government organisation.

Taxation

The principal source of data on taxation is the *Annual Report* of the Revenue Commissioners. This provides an abbreviated description of the tax legislation in force during the year to which the report refers and detailed statistics on the yield from each individual item of taxation. Data are provided on the revenue yielded from:

Wealth Tax (while in operation)
Capital Acquisitions Tax
Corporation Tax
Capital Gains Tax
Customs and excise duties
Death duties
Stamp duties
Income tax
Corporation Profits Tax
Value-Added Tax[17]

In each case the report states the Budget estimate of revenue expected from each source and presents the sums actually raised. Some of these taxes are of only minor importance. The yield from excise duties, income tax, and value-added tax at present accounts for over 90% of total tax revenues.

Excise duties are levied primarily on alcohol, tobacco, vehicles and fuel. They are also levied on betting and bookmaking premises, on firearm certificates and on other types of licences. The yield from the latter sources is trivial. The annual report documents the gross receipts under each heading, any drawbacks or allowances, re-payments and rebates made and a final figure for net receipts to the exchequer. Comparable figures are generally provided in the report for a run of three or four years. Over 30% of all excise duty receipts accrue from duties on hydrocarbon oils, 35% from duties on alcohol, 18% from duties of tobacco, 10% from duties on motor vehicles and their component parts, and the remainder from a variety of other sources. The report indicates in each case whether the duties were levied on imported or home-produced goods.

On the subject of income tax, the annual report documents the allowances and tax bands in force during the year in question, together with comparable figures for earlier tax years. A table is presented of the effective rates of income tax on specimen incomes during the current tax year for single persons, a married couple without children and a married couple with three children. The report documents the amount of gross income declared for tax purposes, the several allowances, reductions and exemptions which were granted, the net yield of tax, the net produce for each penny of the standard rate of tax, and the average effective rate of tax levied on each pound of actual income. These data are presented for a period of five to six years, culminating in the period to which the report refers. Separate data are presented for those taxed under Pay As You Earn (PAYE), and for those paying tax under Schedule E (the self-employed).

Taxpayers are cross-classified by income, sex, marital status, and in the case of married couples, by whether or not the wife is working. The number of tax cases[18] in each cell, the total income accruing to them and the amount of tax paid is documented. Tax cases where allowances have been allotted for children are cross-classified by income range and by the number of dependent children. Such tables are presented separately for those paying tax under Schedule D and under Schedule E. The report also documents the allowances granted in respect of interest, and premiums for life or medical insurance. These allowances are cross-classified by income range, showing in each instance the number of tax cases in respect of which such allowances were granted and the total amount of each type of deduction.

Value-Added Tax (VAT) is a general sales tax levied at all stages of production and distribution. In theory, each liable trader pays tax only on the value he adds to the product in question. This is achieved by allowing him to credit the tax which has been paid on his inputs against the tax payable on his output. VAT is chargeable at different rates on particular types of goods, the report providing a brief resume of the rates in operation during the year in question.

The report documents the number of traders registered for VAT at the outset of the year, the number of new registrations during the year and the number of registrations cancelled. The number of effective registrations at the end of each year is disaggregated by type of trading activity. For each type of activity, the value of goods and services liable to tax is identified, and the amount of deductible inputs and imports; the amount of tax levied on sales and deductible on purchases is shown, together with a net figure of tax paid or refunded in respect of each type of trading.

In analysing the data in the report, it is important to note that different collection periods are in operation for different types of taxes.

The following taxes are levied on a calendar year basis: customs, excise, death, and stamp duties, capital acquisitions tax and VAT. Income tax and capital gains tax are levied on a financial year basis. Corporation Tax and Corporation Profits Tax are levied in accordance with the accounting period of the assessed firm. Thus, while the annual report records the amount remitted under each heading during the period to which the report refers, these taxes will refer to activities in differing periods.

The Returns of Local Taxation document the amounts raised through local taxes, i.e. rates on dwellings and premises. Rates on domestic dwellings were abolished in 1978. For earlier years the Returns of Local Taxation document the total rate revenue received within each local authority area, the rate poundage levied by each local authority, and the product of a penny rate in each area.

Research on taxation in Ireland has concentrated on personal taxation. Bradley (1978) examined the impact of inflation on the burden of personal income taxation during the period 1971 to 1977. Dowling (1977) analysed the income sensitivity of the personal income tax base, and in a later paper (Dowling, 1978) examined the feasibility of introducing an integrated taxation and social welfare system. Norton and O'Donnell (1978), building on Dowling's earlier work, examined the Irish PAYE personal income tax system. O'Muircheartaigh (1977), using time series data over the period 1946 to 1976, has assessed the changing burden of personal income tax in Ireland and the social valuation of income.

Lennan (1972) analysed the built-in flexibility of Irish taxes consequent on changes in the underlying level of economic activity. He focussed on the elasticity of the yield from individual taxes with respect to changes in the level of activity. For the period under study, short- and long-run elasticities in excess of unity were estimated for most important categories of tax. This suggests that the operation of the tax structure in the early part of the 70s had the effect of channelling an increasing proportion of real resources into the government sector. Short-run elasticities for individual taxes were found to be considerably less than unity, ranging from a figure of 0.15 for corporation taxes to 0.29 for personal income tax.

Notes to Chapter 7

1   See *Household Budget Survey,* 1973 Vol. 2, p. 187.
2   *Direct income* includes all money receipts of a recurring nature which accrue directly to the household, together with the value of any free goods regularly received by household members and the retail value of own garden or farm produce consumed by the household before the deduction of taxes or the addition of cash benefits paid by the state. No account is taken of receipts which are of an irregular or non-recurrent nature. The principal exclusions are receipts from sale of possessions, withdrawals from savings, loans obtained or loan repayments received, windfalls, prizes, retirement gratuities, maturing insurance policies, and the like.
3   See *Farm Management Survey Reports* published by An Foras Talúntais, Dublin.
4   See *Redistributive Effects of State Taxes and Benefits on Household Incomes in 1973,* PRL 8628, Stationery Office, Dublin.
5   See Nolan (1978), for example.
6   See the discussion in Nolan's paper.
7   The threshold for a married man was £100,000, for a widow or widower £90,000, and for others £70,000.
8   See the *Irish Statistical Abstract.*
9   These figures are at present unpublished.
10   These data are published in the *Statistical Abstract.*
11   See *Household Budget Survey 1973,* Volume 1, PRL 5415, Stationery Office, Dublin, for a full description of the nationwide survey.
12   See *Household Budget Survey, Annual Inquiry Results for 1978,* PRL 8770, Stationery Office, Dublin, for a full description of the annual urban survey.
13   See *Household Budget Survey 1973,* Volume 1, pp xvii-iii for a discussion of these points.
14   See *Irish Statistical Bulletin* March 1977, pp. 2-8.
15   With effect from 1 January 1978 the rates liability in respect of domestic dwellings, secondary schools, farm buildings (not already derated), and bona fide community halls was terminated. Rating authorities are compensated for the revenue foregone by way of a state grant.
16   For a comprehensive overview of economic aspects of local government in Ireland see de Buitléir, D. (1974) *Problems of Irish local finance,* Institute of Public Administration, Dublin.
17   For years prior to 1972/3, Wholesale and Turnover Tax.
18   The number of tax cases is not synonymous with the number of tax payers. A married couple will be counted as one tax case, except under circumstances where they opt to be treated as two single persons for tax purposes.

8 Prices and Wages

Price Index Numbers

This chapter reviews the availability of data on prices and wages in the agricultural and non-agricultural sectors. It opens with a digression on index numbers which are used in the analysis of prices and wages. The provision of data in the non-agricultural and agricultural sectors is then reviewed.

Index numbers

It is frequently of interest to know how rapidly prices or wages are changing or by what proportion they have increased with reference to some base period. Where only one price, or the wages of one type of labour, is of interest the question can be answered quite simply by means of *price relatives*. Consider an example. Define the price of the good in the base or reference period as P_0 and its price in period 1 as P_1. The *price relative* in period 1 with respect to period 0 is then generally defined as

$$(P_1/P_0)$$

A *price relative* is therefore defined as the ratio of the price of a single commodity in a given period to its price in the base or reference period. It is obvious that the price relative for a given period with respect to the same period is 1.0. In particular the price relative corresponding to the base period is by definition always 1.0.

The price relative defined above may be expressed as a *price index* by multiplying the above ratio by 100, i.e.

$$P_{01} = (p_{01}p_0) \times 100$$

where P_{01} is to be interpreted as the price index for period 1, with base or reference period 0. It is obvious that the value of the index for period 0 will be 100, viz.

$$P_{oo} = (p_0/p_0) \times 100 = 100$$

While the price relative or price index is a simple unambiguous measure of the change in price of a single commodity, matters become more complicated if the aim is to measure the overall or average change in price of a collection of commodities by means of a single measure. Of course, separate price relatives or indexes could be calculated for each commodity, but it is often necessary or desirable to devise a single measure to describe the average change in price of a collection of commodities.

If all prices changed in the same proportion, i.e. if the price relatives for all goods were identical, it would be simple to construct such a price index. For example, if all prices rise by 10% between year 0 and year 1, then the price index for year 1, with base year = 100, would be 110. However, in practice prices do not generally change in the same proportion, and indeed may well move in opposite directions. It is therefore necessary to combine a number of disparate price changes to arrive at a single measure for the overall change in prices.

A simple method of doing so is to calculate the arithmetic average of the individual price relatives for the set of commodities. Consider the case where there are only two commodities. If the price of one commodity increased by 10% between year 0 and year 1 and the price of another by 5%, then their respective price relatives would be expressed as 1.10 and 1.05. The simple average of these price relatives:

$$(1.10 + 1.05)/2 = 1.075$$

yields an index of the overall price change of 107.5. However, this is in general an unsatisfactory way of constructing a price index as it accords equal weight to each commodity. A change of 10% in cattle prices, for example, is clearly of far greater import to the economy than a 10% change in the price of eggs.

In constructing an index it is therefore desirable to *weight* price relatives in some way, in order to reflect the comparative importance of different items. Thus, for example, in constructing an index of consumer prices, it is necessary to take account of the relative *weights* of different commodities in consumer expenditure. Analogously, the weights used in the construction of an index of wage rates should reflect the numerical strengths of different types of labour in the economy. The actual value computed for an index will reflect the weighting pattern chosen, and a number of different methods are available to calculate such prices indexes. The principal methods employed are now described.

One method is to weight individual price relatives by base year value proportions. Algebraically

$$P_{01} = \sum_{i=1}^{n} (p_{i1}/p_{io}) \frac{p_{io} q_{io}}{\sum\limits_{i=1}^{n} p_{io} q_{io}} \tag{1}$$

Here the bracketed expression on the right hand side of the equation is the price relative for commodity i. The following term is a ratio; the numerator is the value (price times quantity) of the output of, or expenditure on, commodity i in the base period (period 0), and the denominator is the value of output of, or expenditure on, all commodities covered by the index in the base period. This ratio therefore reflects the proportional or relative importance of commodity i in the base year. To pursue the simple two-commodity example quoted above, if cattle (commodity 1) comprised 90% of combined output in period 0, and eggs (commodity 2) comprised 10%, a weighted price index for the two commodities in period 1 would be

$$P_{01} = \frac{p_{11}}{p_{1o}} \times 0.90 + \frac{p_{21}}{p_{2o}} \times 0.10$$

where p_{11}/p_{1o} and p_{21}/p_{2o} are the respective price relatives for cattle and eggs. The price relatives have been weighted according to their relative importance in the base year.

An alternative is to weight the price relatives by current year value proportions. This can be expressed as

$$P_{01} = \sum_{i=1}^{n} (p_{i1}/p_{io}) \frac{p_{i1} q_{i1}}{\sum\limits_{i=1}^{n} p_{i1} q_{i1}} \tag{2}$$

The numerator of the ratio here is the value of output of, or expenditure on, commodity i in the current period, while the denominator is the aggregate value of output of, or expenditure on, all the commodities included in the index in the current period. As before, the object is to weight individual items in a way which reflects their relative importance.

Other weighting patterns may be devised, but the two illustrated above are the most common methods of combining price relatives. Over short periods of time (1) and (2) will normally yield similar values. As the time difference between the current and base period lengthens, the values of the two formulae will increasingly diverge. (For a more extensive discussion of this point, see Yeomans (1968)).

The discussion so far has been couched in terms of price relatives. To convert the above formulae to price indexes, it is only necessary to multiply the right-hand side of each formula by 100.

A conceptually different approach to the calculation of price indexes is to bypass the use of price relatives and to compare the cost of producing or purchasing an identical 'basket' of commodities in two periods. For example, define

$$P_{o1} = \frac{\sum_{i=1}^{n} P_{i1} q_{io}}{\sum_{i=1}^{n} P_{io} q_{io}} \times 100 \qquad (3)$$

The denominator of this expression is the aggregate cost of producing or purchasing the given basket of commodities in the base period (period 0) at the prices prevailing in period 0. The numerator is the cost of purchasing the identical set of commodities at period 1 prices. Since the quantities are identical in both periods, any change in the aggregate cost must be due to price changes and the index measures the change in the overall cost of purchasing (or producing) these commodities.

As in the previous examples, the value of the index depends upon the set of weights, in this case the quantities q_o, used. Formula (3) uses as weights the quantities produced or purchased in the base year; this is known as Laspeyres' Price Index formula, and is the method most commonly used in Irish official price statistics. Less commonly used is Paasche's Price Index formula, which uses current period quantity weights, i.e.

$$P_{01} = \frac{\displaystyle\sum_{i=1}^{n} P_{i1} q_{i1}}{\displaystyle\sum_{i=1}^{n} P_{io} q_{i1}} \times 100 \qquad\qquad (4)$$

Here the numerator is the cost of producing or purchasing the current year set of quantities at the prices ruling in that period, while the denominator measures the cost of producing or purchasing the same selection of items at period 0 prices.

Any of the above formulae may be used to compute a series of price indexes for successive periods, the successive indexes being defined in relation to a fixed base period. Laspeyres' price index for period 2, for example, can be defined as

$$P_{02} = \frac{\displaystyle\sum_{i=1}^{n} P_{i2} q_{io}}{\displaystyle\sum_{i=1}^{n} P_{io} q_{io}}$$

In fact, although conceptually distinct, formulae (1) and (3) and formulae (2) and (4) are algebraically identical. Re-writing and simplifying formula (1) (and omitting the subscripts on the summation signs)

$$P_{01} = \sum (P_{i1} / P_{io}) \frac{P_{io} q_{io}}{\sum P_{io} q_{io}}$$

$$= \sum P_{i1} \frac{q_{io}}{\sum P_{io} q_{io}}$$

$$= \frac{\sum P_{i1} q_{io}}{\sum P_{io} q_{io}}$$

which is Laspeyres' Price Index. Similarly, (2) can be shown to be equivalent to Paasche's Price Index.

Neither of these indexes is an unbiased measure of changes over time. Indeed, the Laspeyres index is typically biased upwards — it exaggerates any changes; the Paasche index is typically biased downwards — it underestimates any change. It is extremely difficult to quantify the bias. In the case of the consumer price index, for example, which attempts to measure movements in the overall price of household consumption, the extent of the bias will be greater the more rapidly consumption patterns are changing.

In an attempt to avoid these and other problems, a third index was defined, known as Fisher's Ideal Index. This index is the geometric mean of the Laspeyres and Paasche index numbers given above. This index is now of declining importance in the compilation of Irish economic statistics.

The Non-Agricultural Sector: Prices and Wages

The Consumer Price Index (CPI)

The *Consumer Price Index,* compiled using the Laspeyres formula, is designed to measure changes in the general level of prices actually paid by private households for consumer goods and services. The CPI is therefore inclusive of all indirect taxes. It is calculated on a quarterly basis in respect of the middle Tuesday of the months of February, May, August and November. The CPI and its antecedents stretch back as far as July 1914.[1] The present index is to base November 1975 = 100.

The items included in the present Consumer Price Index, and the weights attached to each item, are based upon the results of the Household Budget Survey, 1973. The present CPI includes 389 items. These items may be aggregated into ten broad headings, for which separate indexes are also published. These separate commodity groups and their weighting in the overall index are as follows.

Item heading	Percentage expenditure weight
Food	30.32
Alcoholic drink	11.39
Tobacco	4.42
Clothing and footwear	10.73
Fuel and light	5.89
Housing	6.06
Household durables	4.82
Other goods	5.18
Transport	13.21
Services and related expenditure	7.99
Total	100.0

(Note that the column does not add up exactly because of rounding).

Price quotations for each item within these commodity groups are obtained from a variety of sources, including shops, the Electricity Supply Board, Coras Iompair Éireann, the Department of Posts and Telegraphs, the Department of the Environment, landlords, principal gas companies, oil companies, garages, doctors and dentists. Prices are generally not adjusted for seasonal variation; however, the prices of eggs, potatoes and tomatoes are seasonally adjusted before being included in the index, as are changes in the cost of accommodation in hotels and guest-houses.

The CSO stresses that the CPI is a price index and *not* a cost of living index. As a Laspeyres index, it is specifically designed not to take into account the manner in which households change their pattern of expenditure in response to changes in prices, incomes, family composition, tastes or market conditions. Furthermore, items such as income tax and social insurance contributions which have an important effect on household budgets are not included in the CPI.

The CPI is not a perfect measure of changes in the general level of prices experienced by any particular household. All households differ, and no household is likely to have an expenditure pattern identical to that implied by the weights in the index. In order to overcome this problem some countries, but not Ireland, produce separate consumer price indexes for different types of household. Kennedy and Bruton (1975) performed a disaggregated analysis of household expenditure by social class over the period 1968 to 1975 and concluded that the officially published consumer price index was appropriate for all household types.

The Consumer Price Index is published quarterly in the *Irish Statistical Bulletin.* As a by-product of the compilation of the index the Bulletin also includes a table listing comparable national average retail prices of some food products. Forty-four individual products are distinguished and their national average unit prices given for the present and preceding quarter.

In November 1976 a *Constant Tax Price Index*[2] was introduced. Since indirect taxes and duties are levied on many items included in the Consumer Price Index, changes in such tax rates will lead to a change in the index. Furthermore, the impact on the index of increases in the prices of goods subject to proportional taxes will be magnified by the increased absolute amount of tax levied. The constant tax price index attempts to exclude both these effects. The method of computation adopted keeps constant in money terms the mid-November 1975 indirect tax content of the basket of goods priced quarterly for compilation of the Consumer Price Index. Rates on dwellings and motor taxation are also held constant at end 1975 levels. The index therefore represents the percentage change between the cost of the index basket at mid-November 1975 prices and its costs in a subsequent period, less the increase in the indirect tax content during that period.

The government hoped to preserve flexibility of budgetary action by promoting this index in pay bargaining negotiations. Their underlying reasoning was as follows. In cases where wage increases are linked to an index of consumer prices, changes in indirect tax rates can have significant inflationary effects. If instead wages are linked to an index which excludes the effect of such tax changes, the inflationary effect should be considerably less. The government's aim has not been achieved, however, and the index has attracted little attention from either employers or employees.[3]

The Wholesale Price Index

The present *Wholesale Price Index* to base 1975 = 100 was introduced in March 1978.[4] It aims to provide monthly output price indexes for each sector of manufacturing industry.[5] The index is based on approximately 3,300 monthly price quotations provided by 460 respondents in the manufacturing and wholesale sector. The results are published quarterly in the Irish Statistical Bulletin.

Wholesale price indexes are calculated for four broad classifications:

1   Output of manufacturing industry
2   Building and construction materials
3   Capital goods
4   General wholesale price index

In each case the indexes are compiled using the Laspeyres index formula.

The manufacturing industry wholesale price indexes distinguish twenty-four major sectors of manufacturing industry at the two-digit NACE level. The indexes for each of these sectors are calculated as a weighted average of the monthly price relatives for constituent commodities, the weights being the net selling value of the output of these commodities as returned in the 1973 Census of Industrial Production. The weighted average of the indexes for individual industries is then calculated to produce an overall wholesale price index for the output of manufacturing industry. The weights used in this latter calculation are net sector output value weights estimated from the 1969 input/output table.

In the building and construction sector separate indexes are published for eleven categories of commodity, as well as an overall index for all building and construction materials combined. The disaggregated indexes are generally derived as the simple arithmetic mean of the monthly percentage change in the prices of the constituent items surveyed. The overall index is then calculated as a weighted average of these disaggregated indexes, the weights being derived from the 1973 *Census of Building and Construction.*

The wholesale price indexes for capital goods distinguish those employed in transportable goods industries and those employed in building and construction. Separate indexes for transportable capital goods are presented for agriculture, industry and other. Within the industrial sector separate indexes are published for private vehicles and commercial vehicles. The building and construction capital goods wholesale price index is calculated by combining a special hourly wage rate index for employees in the building and construction sector with the price index for building and construction materials described above. The overall wholesale price index for capital goods is a weighted average of these disaggregated indexes, the weights being based on the distribution of fixed capital formation in the 1974 national accounts.

A *General Wholesale Price Index* is also published. This is calculated from the price indexes for the Output of Manufacturing Industry, Output of Agriculture (see below) and Import Unit Value Price Index (see below), modified as far as possible to minimise duplication. The weights used to combine these indexes to produce the General Wholesale Price Index were estimated from the 1969 Input/Output Table.

The Wholesale Price Index Series introduced in 1978 replaced an earlier series to base 1953 = 100. In later years, this original series had become of limited use because of the dated weighting pattern it employed. To prevent the recurrence of this problem new firms will be canvassed for co-operation and the range of products priced from existing respondents will be revised on a regular basis. The weighting basis of the new index will also be regularly revised.

The inadequacies of the older Wholesale Price Index in later years caution against linking the two series. There is a further incompatibility between the two resulting from the introduction of VAT as a replacement for turnover and wholesale taxes in 1972. The new series is based on prices exclusive of invoiced VAT. The older series by contrast included VAT in those flows in respect of which the users could not get a refund.

Import and export unit values and the terms of trade

In the case of traded goods *unit value indexes* are computed instead of price indexes. 'Unit values' are obtained by dividing the total value of a group of commodity imports (exports) by the total quantity of these imports (exports). To the extent that each group of commodities is homogeneous — i.e. the units comprising each group are physically identical — 'unit values' are synonymous with prices. In practice most commodity headings comprise a mixture of similar although not identical commodities. The unit values calculated for each commodity heading are therefore a kind of crude average price for a number of different commodities. While this qualification should be borne in mind it is reasonable to interpret the unit values as prices for most purposes.

The method of construction of these indexes is quite complex. In particular, different methods are used to produce the annual and monthly indexes. The monthly indexes are computed using the Laspeyres formula, while the annual employs Fisher's 'ideal' formula. The unit value index for a particular month is computed as follows.

The denominator is the aggregate cost of a selected basket of imports (exports) in the *previous year* valued at the average unit values prevailing in that year. The numerator is a measure of the cost of this same group of imports (exports) at the prices prevailing in the present month. The result is a price index for the current month which is related to the average price level in the previous year. The value of the index may then be converted to some fixed base year to form part of a continuous series. (This technique is called the 'chain-link' method; for a discussion see Chapter 4.) Monthly import and export unit value indexes for Ireland are currently to base 1975 = 100. Thus the monthly price indexes for 1981 are first calculated in relation to the average level of prices in 1980 and are subsequently expressed in relation to the average level of prices in 1975.

The monthly index takes no account of the quantities imported or exported in the month for which the index is calculated. The quantities included in the index relate solely to the relative quantities imported or exported in the previous year; they may not reflect accurately the actual pattern of trade during the month for which the index is calculated. The value of the monthly index is therefore independent of the actual pattern or volume of trade during that month.

This treatment is justifiable on several grounds, one of which is that the principal object of the monthly unit value index is to indicate the trend in the annual index: it is therefore logical to use annual quantities of imports or exports as weights and ignore seasonal variation in the pattern of trade. Moreover, by using the same set of quantities as weights for each month, month-to-month changes in the value of the index are directly comparable. This would not be the case if the weights changed from month to month, as the value of the index would then fluctuate as a result not only of changes in prices but also because of seasonal variation in the pattern of trade. Seasonality in the pattern of trade is thus not a factor in these measures of price change, while the seasonality inherent in the prices themselves has not been significant.

The annual unit value indexes for exports and imports differ from the monthly indexes both in method of calculation and in coverage. With respect to coverage, both the annual and monthly indexes are based upon a representative selection of commodities, but the selection used for the annual index is more comprehensive than that used for the monthly one. Furthermore, whereas the monthly index is computed using the Laspeyres formula, Fisher's ideal index formula is used to calculate the annual index. This can be written

$$P_{01} = \sqrt{\frac{\Sigma \, P_1 \, q_0}{\Sigma \, P_0 \, q_0} \times \frac{\Sigma \, P_1 \, q_1}{\Sigma \, P_0 \, q_1}} \times 100$$

where P_{01} is the annual price index number for year 1, with the base period year 0 equal to 100.

$p_0 q_0$ is the aggregate cost of a selected group of imports (exports) in period 0, at period 0 prices.

$p_1 q_0$ is the aggregate cost of these same imports (exports) at period 1 prices.

$p_1 q_1$ is the aggregate cost of a selected group of imports (exports) in period 1 at period 1 prices.

$p_0 q_1$ is the aggregate cost of these same imports (exports) at period 0 prices.

The first ratio under the square root sign will be recognised as a

Laspeyres price index (base period quantities are used as weights); the second is a Paasche price index (current period quantities are used as weights). Generally these two indexes will be different in value and the 'actual' annual price index is taken as the geometric mean of the two.

In the present context, the period 1 prices and quantities may be taken to represent the average annual prices and quantities of a selected group of imports (exports) in the current year, while the period 0 prices and quantities refer to the average prices and quantities of the same group of imports (exports) in the previous year. This gives rise to an index for the current year which is based on the average level of prices in the previous year. The value of the index may then be converted to a fixed base year to form a continuous series.

Both the annual and the monthly price indexes are based upon a representative sample of imports and exports and as such are estimates of changes in the price levels for all imports and exports. Because of the range of commodities included and the particular price index number formula used, the annual price index may be claimed to reflect more accurately the overall trend in import and export prices. By virtue of the independent method by which the annual index numbers are calculated, they will not generally equal the mean of the twelve monthly indexes. The different methods of calculation and the fact that the annual index numbers are based on more items accounts for the fact that the annual unit value indexes may differ fairly substantially from the average of the monthly figures. The CSO caution that the monthly figures should be regarded as of an interim nature and as suitable only for month-to-month comparisons within a given year.

The *terms of trade* is an expression used to describe a comparison of relative changes in import and export prices. Changes in such prices are of interest in relation to a country's international competitiveness and its balance of payments. The object of a measure of the terms of trade is to determine the extent to which such relative price movements may be said to have an adverse or a favourable effect for the country concerned.

A rise in the import price index implies that imports have become more expensive relative to some base period. If export prices have remained unchanged, or have fallen, or if the export price index has risen by less than the import price index over the same period, this means that a greater volume of exports is required to purchase the same volume of imports, relative to the purchasing power of exports in the base period. In this case there has been an *adverse* movement in the terms of trade. Conversely, where there has been a rise in the export price index which surpasses that in the import price index, the resulting movement in the terms of trade is described as *favourable*. A relative rise in export prices is favourable to the exporting countries, in that it raises the real purchasing power of exports, whilst a relative rise in import prices implies that a greater volume of domestic output must be exported to pay for the same volume of imports.

Changes in the terms of trade are expressed numerically as the ratio of the export unit value index to the import unit value index. For any period T the 'terms of trade index' may be calculated as

$$\frac{\text{Export Unit Value index for period T}}{\text{Import Unit Value index for period T}} \times 100$$

If export prices have risen relative to import prices the terms of trade index will be greater than 100, implying a favourable movement in the terms of trade relative to the base period. If import prices have risen relative to export prices the terms of trade index will be less than 100 and this implies an adverse movement in the terms of trade. Changes in the index from one period to another show whether the terms of trade have 'improved' or 'deteriorated'. Note that the choice of base year determines both the magnitude and the direction of change in the index so changes in the terms of trade index must be interpreted in relation to the base period selected.

The terms of trade index may be interpreted in the following way. The volume of imports which can be purchased in exchange for a fixed quantity of exports depends on the relative prices of both imports and exports. The index may therefore be interpreted as measuring the volume of imports which can be purchased in each period in exchange for the base period volume of exports, subject to the prices prevailing in each period. The volume of imports which can be purchased in the base period is taken as 100. If, between the base period and the current period export prices have risen relative to import prices, it is clear that a greater volume of imports can be purchased in exchange for the same volume of exports. Under such circumstances the terms of trade may be said to have 'improved'.

In practice, in the world of fixed exchange rates prevailing up to 1973, changes in the terms of trade must be interpreted cautiously. A relative rise in export prices implies a favourable movement in the terms of trade index but may result in a fall in the volume of exports; this depends upon the price elasticity of demand for such goods. This may cause a deterioration in the exporting countries' external trading position since the new level of export receipts may be insufficient to maintain the previous level of imports. Conversely, increases in productivity may result in a fall in export prices and a greatly increased volume of exports; under such circumstances the fact that there has been a deterioration in the terms of trade may be somewhat misleading. In interpreting this index its limitations should therefore be borne in mind.

Indexes of the terms of trade and of the unit value of imports and exports are recorded in the Irish Statistical Bulletin and in the *Annual Statistical Abstract.* Unit value indexes are included in the monthly release of provisional trade figures and in the monthly *Trade Statistics of Ireland* (see Chapter 5). The current base year for these indexes is 1975 = 100, so that they measure relative changes in import and export prices and in the terms of trade in relation to their respective levels in 1975.

House prices
The Department of the Environment publishes quarterly data on both new and second-hand house prices. These series are published in the *Quarterly Bulletin of Housing Statistics.* In the case of new houses the series documents the average gross price of those houses for which loans were approved by the following types of lending agency:

> building societies
> insurance companies
> local authorities and
> all agencies

The figures are given nationally and for six sub-national areas, namely the counties of Cork, Dublin, Limerick, Waterford and Galway, and for other areas outside these counties. Similar figures have been provided for second-hand houses on a quarterly basis since 1975. The interpretation of trends in both series is somewhat difficult; neither have a uniform weighting system across the three-month periods by type of house or by type of lending agency. The series simply records the average price of all relevant transactions notified during the quarter in question.

The *Quarterly Bulletin of Housing Statistics* also includes an *Index of House Building Costs.* This index, to base January 1975 = 100, is prepared on a monthly basis by Dublin Corporation and relates solely to the labour and material costs of housebuilding. Its coverage is therefore rather narrow in that it does not include items such as builders' overheads, their profit margins, their interest charges or the costs of land and development.[6]

Two further prices indexes relevant to the construction sector are to be found in *Construction Industry Statistics* published annually by An Foras Forbartha. The first is the 'National Building Price Index', published quarterly to base January 1975 = 100. Disaggregated indexes are published for five separate building types. The second is the quarterly index compiled since 1966 by the Irish Branch of the Royal Institute of Chartered Surveyors. Separate indexes are published for construction costs and labour costs. Some of the information contained therein is reproduced in more condensed form in the *Annual Report* of the Department of the Environment.

A series on the *Average Weekly Rents of Dwellings Let by Local Authorities* is also published in the Quarterly Bulletin of Housing Statistics. The series is annual, the data for years up to 1973 being by financial year. Dwellings let by County Borough Corporations, Urban District Councils and County Councils are separately distinguished.

Financial prices
The CSO has compiled an *Index of the Prices of Ordinary Stocks and Shares* on a monthly basis since 1953. The index is based on official quotations of share prices at the beginning of each month. Since 1967 only the shares of companies with market capitalisation exceeding £500,000 have been included. A series is also published on a monthly basis of the average price of a representative short-term government stock. Both series are published in the *Economic Series* section of the Irish Statistical Bulletin.

The cost of capital
While the cost of employing labour in the production process is relatively well documented, the cost of employing the other productive factor, capital, is scarcely documented at all. No official series exist on the cost of capital to Irish industry. Attempts have been made by Geary, Walsh, and Copeland (1975) and by Geary and McDonnell (1979) to fill this gap. The latter source estimates the cost of capital to Irish industry under various assumptions about interest rates, scrapping rates, depreciation and the hiring of capital services. The study also attempts to take account of the subsidy to capital costs inherent in IDA

grants. All measures computed suggest that labour costs in Ireland are tending to increase more rapidly than those of capital.

Wages

It is important to distinguish between wage rates and wages (or earnings). The former reflects the contractual bargain struck between employer and employee while the latter also reflects the average length of the work week. Earnings therefore are a combination of wage rates and hours of work. The relationship between the three concepts is not a simple multiplicative one, as marginal hours may be remunerated at overtime rates. In analysing data on the cost of labour services the particular purpose for which the analysis is being undertaken will determine whether earnings or wage rate data are the more appropriate.[7]

Wages in transportable goods industries and in the building and construction sector are relatively well documented.[8] By contrast, little information is available on wages in the tertiary sector. The main data source in the industrial sector is the *Quarterly Industrial Inquiry*.[9] This inquiry uses the same sample of firms as does the Monthly Industrial Inquiry described in Chapter 4. The results of the inquiry provide quarterly data on average weekly and hourly earnings of employees at the two digit NACE industry level. The results separately distinguish men and women employed at adult rates and all industrial workers. The results are published in the Irish Statistical Bulletin. The annual Census of Industrial Production also provides data on the total amount paid in salaries and in wages and earnings in each industry during the year in question.

Since 1969 a *Quarterly Inquiry on Earnings and Hours Worked in the Private Building and Construction Industry* has been carried out. The results detail average weekly and hourly earnings for several categories of staff, including male and female clerical staff, wage-earners paid on a time basis (separately distinguishing foremen and supervisors, skilled operators and apprentices), and unskilled and semi-skilled operators (distinguishing those employed on adult rates and those employed on other rates). In interpreting the results of this enquiry it is important to bear in mind that the earnings in any quarter will reflect the weather in that quarter. Bad weather will significantly depress average earnings.

A considerable amount of data is published on an annual basis in the Irish Statistical Bulletin on the average weekly wage (or salary) and the average earnings of different grades of staff employed on the railway operations of Coras Iompair Éireann. The series is a quarterly one, being based on a particular week within each quarter.

The agricultural sector

Agricultural prices

A vast amount of data exists on agricultural prices in Ireland. Indeed, if the output of other sectors was documented in such excruciating detail the country would be awash with price statistics. New price indexes for agricultural input and agricultural output were introduced on a harmonised EEC basis in 1975.[10] These will be considered in turn.

The *Agricultural Input Price Index* to base January 1975 = 100 replaced the old Price Index of Certain Farm Materials Purchased at Retail Prices. The aim of the revised index is to measure changes in the prices of all farm inputs purchased for current consumption by the agricultural sector. The all-input index can be disaggregated into a variety of sub-indexes. These include:

a   all feeding stuffs and three sub-categories thereof
b   all fertilisers and three sub-categories thereof
c   seeds
d   energy separately distinguishing motor fuels and electricity
e   other inputs of a non-capital nature

The indexes are compiled on a monthly basis using the Laspeyres index formula. The use of a Laspeyres index marks a departure from earlier practice, as the superseded series was computed using Fisher's ideal index formula. The weights used in compiling the new index are derived from estimated expenditure on individual farm inputs averaged over the three years 1974, 1975 and 1976. It is intended to re-base the series at regular intervals in order to keep pace with changes in the pattern of agricultural inputs over time. In the case of inputs for which monthly data on purchases during the base year are available, for example on feeding stuffs and fertilisers, the annual index numbers will be compiled independently of the monthly series. The annual indexes for such inputs will be obtained as weighted averages (using 1975 monthly consumption weights) of the monthly prices. In such cases the annual index numbers will generally not be identical with the simple arithmetic average of the monthly figures. In the case of those inputs for which monthly, quarterly or half-yearly data on purchases during 1975 are not available, the annual index numbers will be obtained as the simple averages of the monthly indexes. An important point to note is that where applicable, VAT is included in the prices used to compile the index numbers. This contrasts with the compilation of the non-agricultural Wholesale Price Index which excludes VAT.

The *Agricultural Output Price Indexes* are intended to measure trends in the price levels of agricultural produce sold by farmers. The construction of this index is very similar to that of the agricultural input index documented above. The index is a Laspeyres one; the weights are the estimated value of sales of different types of produce off farms (excluding inter-farm sales and value of own produce consumed by farm households) averaged over the three years 1974, 1975 and 1976. The use of a Laspeyres index represents a departure from the procedure used to compile the superseded 1953-based series of index numbers. The caveats made about comparing monthly and annual values of the agricultural input price indexes are equally relevant here.

The new index is considerably more detailed than the old one. It documents output prices for livestock (separately distinguishing four categories), for livestock products (separately identifying milk), for livestock and livestock products combined, and for crops.

In addition to these global indexes, a wide variety of disaggregated secondary data is available on prices in the agricultural sector. The Irish Statistical Bulletin contains monthly data on the average price per head of bullocks, heifers, sheep and young pigs in certain weight classes at livestock auction marts. The series on sheep prices was inaugurated in 1975; the series on other categories of livestock extends back several decades. Prices of individual farm inputs are also documented, including the average monthly and quarterly retail prices of feeding stuffs and the annual average retail prices of seeds and fertilisers in towns.

Data are also published in the *Irish Statistical Bulletin* on the average annual prices of individual agricultural products. These include potatoes, eggs, porkers, bacon pigs, wheat, oats, barley, hay, chickens and turkeys. The figures record an aggregate price for all markets, for the Dublin market on its own and for all markets excluding Dublin.

New data on milk prices have been collected following Ireland's entry to the EEC. These record the monthly prices paid by creameries per gallon to milk producers; prices vary according to butterfat content and a standard product has 3.7% butterfat. This has been collected and published in the Irish Statistical Bulletin since 1975.

Agricultural wages
The main data source on the wages and earnings of agricultural

employees is the annual *Survey on the Earnings of Permanent Male Agricultural Workers* introduced in 1975[12] as part of an EEC-wide operation. Data are collected from a sample of holdings reporting permanent male employees in the June agricultural enumeration. The information elicited includes the sex of the employee, age, gross earnings, number of hours paid for, the basis on which the earnings are calculated, any non-wage benefits, the type of work, and the occupational status of the employee. The results are published in the Irish Statistical Bulletin.

The *General Index of Wage Rates* (base 1953 = 100), published in the Irish Statistical Bulletin, is of longer standing. This series includes an index of agricultural weekly earnings in July of each year. However, the series is of limited value as the figures are based on the average minimum weekly wage rates payable to adult male agricultural workers under the *Agricultural Wages (Minimum Rates) Orders* of each year.

The amount of wages paid in forestry operations is also published in the Statistical Bulletin. This includes the monthly wage bill paid and the average number employed each month. A figure for average earnings in forestry can be derived from this.

Significant improvements have been made in the provision of data on prices. The introduction of the revised Wholesale Price Index and the new Agricultural Input and Output Price series marks a considerable step forward. By contrast, on the wages front, there have been few innovations in the past decade. Perhaps some of the resources allocated to the production of multitudinous individual agricultural prices series would be better employed in improving the quality of wage rate and earnings data, particularly in the tertiary sector.

Notes to Chapter 8

1   See *Irish Statistical Bulletin* March 1976 for details of present and past series.
2   An earlier exercise of this nature was carried out by Baker and Neary (1971).
3   The formation of price expectations in Ireland has been explored by Conniffe and Killen (1977), while the transmission of inflation via exchange rates has been examined by Geary (1976a, 1976b).
4   See *Irish Statistical Bulletin* March 1978, pp. 4-9.
5   For an analysis of trends in earlier series see Black, Simpson and Slattery (1970).
6   Trends in land prices have been described by Chambers (1974).
7   See for example Whelan and Walsh (1977).
8   See Cowling (1966) and Nevin (1962B) for an analysis of historical trends.
9   See Chapter 4 above.
10 See *Irish Statistical Bulletin* March 1979, p. 5 and March 1978, pp. 2-3 respectively.
11 See *Irish Statistical Bulletin* March 1978, p. 76.
12 See *Irish Statistical Bulletin* March 1977.

9 Distribution, Transport and Communication

Distribution

The first Census of Distribution in Ireland was taken for the calendar year 1933. Compared with subsequent enquiries, it was limited in scope; it covered only retail trade outlets and a limited number of wholesale trade establishments. Service trades were completely excluded.

It gave information on the number of persons engaged, wages and salaries paid, the value of sales, the value of stocks at the beginning and end of the year, the class of business and details of ownership. Many of these data were broken down by geographical area in the results. However, no information was collected on purchases, so that no estimates could be made of gross margins in retail and wholesale trade.

The next census of distribution, taken in 1951, was a good deal wider in scope. A number of service trades were included, mainly restaurants, cafes, hotels, guesthouses, cinemas, theatres, hairdressers, pawnbrokers and bookmakers. The coverage of the wholesale trade was also more comprehensive. Information was collected on sales, principal commodities sold and services provided, employment, wages and salaries, stocks and other supplementary information. An important addition was data on purchases; this enabled estimation of gross margins. The published results of the 1951 census included analyses of the geographical pattern of trade and sales, sales per person engaged and per head of population, employment, wages, salaries, turnover and gross margins per person engaged and as a percentage of sales. The results were further classified by type of business. The report included a detailed analysis of distributive trade in towns with a population of 5,000 and over.

The 1951 census was followed by sample inquiries for each of the years 1952-55; the sample statistics were grossed up to provide estimates for all establishments. Each inquiry collected data on sales and purchases, stocks, wages, salaries and employment. In addition the 1953 and 1954 inquiries collected information on capital investment.

The third full census of distribution in 1956 was similar in scope and coverage to that of 1951. Subsequently, sample inquiries were conducted for each of the years 1957-60. The sample inquiries of 1957-60 were smaller in size than those of 1952-5 and differed slightly in coverage: data were collected on sales and purchases, stock changes and gross margins for different descriptions of business, but the published statistics did not include estimates of numbers engaged or of wages and salaries. In addition to retail and wholesale establishments, the inquiries also covered hotels, guesthouses and cinemas. In 1958 and 1959 information was collected on advertising, including expenditure on advertising by retailers and wholesalers, and on the business handled by Irish advertising agencies.

The 1966 *Census of Distribution and Services* covered all permanent business establishments engaged in retail or wholesale trade, as well as those involved in the provision of certain services, namely: catering, hairdressing, shoe repairing, television and radio rental, vehicle hire, auctioneers, estate agents, bookmakers, cinemas and theatres.[1] The information sought from respondents was wide-ranging.

It included the following:

description of the business
list of the principal commodities sold or services provided
turnover or sales during the year (distinguishing wholesale, retail
    and service receipts)
value of purchases
value of stocks at the beginning and end of the year
wages and salaries
numbers employed in a particular week
legal identity of the undertaking (partnership, limited company,
    etc)
membership of wholesaler or retailer trading groups
extent of self-service or cash and carry business undertaken
selling space in square feet
receipts from hire-purchase sales,
debtors and creditors outstanding at the end of the year
number of other premises at which the owner carried on a
    business

expenditure on advertising and transport
changes during the year in fixed capital assets

During the period 1966-71, an annual inquiry into retail trade and a similar inquiry into wholesale trade were carried out by the CSO. The results were published in the Irish Statistical Bulletin. In the case of retail trade, twelve categories of business were distinguished; the inquiry documented the annual percentage changes in sales, purchases, end of year stocks and gross margins for each. Two derived ratios were also presented: gross margin as a percentage of total sales and the rate of stock turnover. Similar data were provided for nine categories of wholesale trade. In both inquiries, data were elicited from a sample of respondents to the 1966 Census of Distribution and Services.

The *Census of Distribution 1971* sought almost identical information to that of 1966, with the significant addition of a question on the duration of the present ownership of the undertaking. The data sought related in principle to the calendar year 1971 though figures for the nearest accounting year were accepted. The replies covered an estimated 83% of retail and 85% of wholesale sales. Estimates for non-respondents were calculated on the basis of returns made by comparable establishments.[2] Certain tables refer to respondents only while others include estimates for non-respondents. Only the latter type of results are comparable between censuses.[3]

The 1971 inquiry covered a smaller range of services than in 1966[4] but its results are more detailed. The 1971 results distinguish thirty-two types of retail undertaking and seventeen types of wholesale undertaking. Separate volumes of results were published for wholesale trade and services and for retail trade. In both cases there is a significant degree of areal disaggregation of the results: distinguishing the state as a whole, the provinces, Dublin and Dun Laoghaire (including their suburbs) and each of the other three county boroughs (including suburbs). The basic data provided for each class of activity include the following:

Number of establishments
Annual turnover
Annual purchases
Opening and closing stocks
Gross margin

Annual wages and salaries
Selling space
Number of persons engaged (proprietors, unpaid family members
    or paid employees)

In addition, these basic data are cross-classified by turnover range and by the number of persons engaged.

The results for service establishments follow the same broad lines as those for wholesale and retail trade, with the addition of data on the legal status of the establishment and on the total number of service establishments operated in the state by the enterprise. The results for the services sector for 1971 and earlier censuses have been analysed by Cogan (1978).

The *Census of Distribution 1977* was restricted solely to wholesale and retail trade. Unlike the 1971 Census, a detailed breakdown of turnover was not requested, a global figure for total turnover only being sought. Otherwise the range of questions included broadly corresponded to those of 1971.

The first annual sample survey of distribution is planned for 1982. The survey will seek information on stocks with a view to monitoring the inventory cycle. Changes in stock levels typically account for the initial stages of both recession and recovery in aggregate demand — yet data on them is almost non-existent. For an attempt to compile such data see Geary (1973). McCarthy (1977B) has specified and estimated an econometric model of non-agricultural stock changes using data from the National Accounts.

The CSO carries out an annual inquiry into the *business of advertising agencies*. The results document the number of persons engaged in April of each year and distinguish working proprietors and full-time and part-time paid employees. Other data secured include the gross amount charged to clients, the categories of advertising space or production work on which this money was spent, the amount spent on market research and total expenditure on wages and salaries. The results are published in the Statistical Bulletin.

Transport

Available statistics on transport may be divided into two categories: those describing the state and extent of the transport network and those documenting the use that is made of it. The first category is particularly relevant to the road and rail networks.

The road network is described in considerable detail in the *Annual Inventory of Public Roads in Ireland* published by the Department of the Environment. The inventory is compiled from data supplied by each local authority on the roads within its administrative area. This publication classifies the roads in each county at year end by type of road; it distinguishes national primary routes, national secondary routes, main roads, county borough roads, county roads and urban roads. Primary routes are the most important inter-urban routes and comprise some 3% of the network. Secondary routes, which comprise a similar percentage, consist of the most important links between the primary routes. Main roads, sometimes called trunk or link roads, represent about 12% of the network. With the passage of time, sections of such roads are being ungraded to primary or secondary standard. County borough roads comprise those within the boundaries of the county boroughs of Cork, Dublin, Limerick and Waterford. Roads in the seven boroughs and the forty-nine urban districts not classified as primary, secondary or main roads are titled urban roads. Roads which lie without the four categories described above are called county roads. The aggregate length of such county roads accounts for over 80% of the total public road network. The roads under the control of each individual local authority are further classified by type of surface, three types being distinguished; bituminous, cement/concrete, and gravel/broken stone. A summary of the data is included in the Annual Statistical Abstract.

The Statistical Abstract is also the major source of data on the mileage of railway track open for traffic on 31 December of each year. The tables presented also identify those stretches which have been reduced to single track, and the mileage of those sidings similarly reduced.

Turning now to the use made of the road network, the CSO publishes monthly returns on the number of vehicles registered for the first time in each county by size of engine and vehicle type. Detailed summaries are included in the Irish Statistical Bulletin and the Statistical Abstract. The Department of the Environment also documents the number of vehicles under current licence in each county by size of engine and vehicle type and the number of driving licences current in each county. Computerisation of the vehicle registration system was completed in 1979; additional data on the structure of the national vehicle fleet will become available in the future.

Information on the intensity of use of the national vehicle fleet and the national road network is available from a wide range of sources. The Central Statistics Office carried out a sample survey of road freight traffic in 1964.[5] This was the first comprehensive study of road freight transport in Ireland, and provided the first estimates of the extent of own account haulage. The survey documented the tonnage carried by road freight, the miles travelled and, as a measure of the industry's output, the number of ton miles. This sample survey of 1964 provided the basis for a *Continuous Sample Survey of Freight Transportation* initiated in 1980 by the Central Statistics Office. This survey is designed to collect data on aspects of road freight transportation including tonnes, mileage and type of commodity carried, type of vehicle, fuel consumed and number of trips. Survey results will distinguish the type of fuel used, and average fuel consumption. Amongst the trip data sought are the number of trips, their origin and destination, the amount of freight carried and the amount of freight collected or deposited at ports and railheads. The vehicle data sought will include the weight, capacity, age and body type of the vehicle. Further data are sought on axle numbers and on the use of trailers and containers. The first results of this survey are to be found in the June 1981 issue of the Statistical Bulletin.

Before the introduction of this survey, the Statistical Abstract and the Irish Statistical Bulletin contained data on *road freight transport by licensed hauliers*. The data were based on an annual survey of the activities of licensed hauliers (including CIE) carried out by the CSO since 1953. The results relate only to carriage of freight for reward — own account haulage was therefore excluded. It also excluded certain categories of freight carried within a fifteen-mile radius of Dublin and Cork or a ten-mile radius of Limerick, Waterford and Galway for which a licence was not required. The data included the number of operators, the number of vehicles, the total vehicle miles run, the tonnage carried and the total receipts of the operators. The results were cross-classified by fleet size (number of vehicles), and were published annually.[6]

Whereas the data referred to above relate only to freight transport, estimates of overall traffic flows on national routes have been published by An Foras Forbartha. Their estimates for 1977 identify all significant sections of road on the national road network and assign to each a current best estimate of the twenty-four-hour traffic flow on an average day during the year. The average heavy goods vehicle content of traffic on each stretch during normal daylight hours is also estimated.[7] An Foras have also published estimates of vehicle kilometres of travel for the year 1979.[8] The estimates contained in this publication are based on the results of a five-year traffic-counting programme conducted by An Foras and the Department of the Environment since 1975. The estimates are compiled by combining hourly travel patterns from self-recording permanent traffic stations with data from visual classification counts. The results provide estimates of the gross travel by each type of road vehicle on the various classes of public road, national, regional, county and urban.

Road accidents are the single greatest cause of death for those aged under thirty-five in Ireland. The number of such accidents recorded by the Garda Síochána are documented in An Foras Forbartha's annual publication entitled *Road Accident Facts.* This publication itemises the numbers killed and injured by the month, day, time and county of occurrence. Accidents are also cross-classified by the light and weather conditions prevailing at the time, by the type of road surface on which the accident took place and by the age and sex of the road-user involved. The Department of Transport publishes the results of inquiries into individual rail and air accidents when these occur.

Statistics of *omnibus passenger road services* are published in the Irish Statistical Bulletin and in the Statistical Abstract. These statistics document for each four-weekly period throughout the year the number of vehicle miles run, numbers of passengers carried and the gross receipts from passenger fares. The statistics refer only to scheduled services. The Statistical Abstract separately distinguishes Dublin city and Dublin suburban services, other city and town services, internal and cross-border services. The Bulletin also contains an *Annual Review of Passenger Road Services,* documenting vehicle miles run, number of passengers carried and the gross receipts of carriers.

Details of CIE rail traffic are published in the Statistical Abstract. In the case of passengers carried, three classes are distinguished: first class, standard class and season-ticket holders. The tonnage of freight traffic carried and the number and type of livestock transported in this manner are documented. A separate table analyses the principal commodities conveyed by rail, showing the annual tonnage carried of twenty-three principal categories. The same source also lists the *revenue receipts and expenditure on railway working by CIE.* Somewhat more aggregated results have been published on a regular basis in the Irish Statistical Bulletin since 1963. This source lists in respect of each four-weekly period the volume of receipts accruing to CIE from passengers, parcels and mail and other receipts from merchandise traffic. The number of passengers carried in each period is documented, as is the tonnage of goods, distinguishing general merchandise and minerals. Annual *statistics of port traffic* are to be found in the Irish Statistical Bulletin and in somewhat more aggregated form in the Statistical Abstract. These include the number of trading and passenger vessels arriving and departing at Irish ports, their net registered tonnage and the number of livestock handled. Foreign registered vessels are separately identified. In the Statistical Bulletin such returns are provided for individual ports. The tonnage of goods handled is also recorded, distinguishing between that received and that forwarded. Separate data are provided for cross-channel and other foreign and coastal trade. The Statistical Bulletin also provides detailed information on unitised traffic at ports. The number of units received or forwarded and the tonnage of unitised goods[9] thus handled at each port is recorded as is the number of empty units. Separate data are also published for roll-on, roll-off traffic. Vehicles travelling in this manner are classified as accompanied private cars or buses, vehicles for export or import and goods vehicles or trailers. The tonnage carried in this manner is documented, as is the number of empty vehicles or trailers thus transported.

The numbers and tonnage of vehicles on the Irish Shipping Register are published in the Statistical Abstract. Three classes of vessel are distinguished: motor vessels, sailing vessels and fishing vessels. The Abstract also documents shipping casualties off the coast, identifying the numbers of lives and vessels lost or salvaged. The activities of Irish Shipping Ltd, a state-sponsored body, are described in some detail. The basic information includes the numbers of staff employed, the number of ships currently operated and their registered dead weight.

Data on air traffic into and out of Ireland is available in the Statistical Bulletin. The number of passengers and the amount of freight and mail travelling into and out of Ireland by this mode are documented monthly for each airport. The Bulletin also records monthly *passenger movement by sea* and *cross border passenger movement.* The former series distinguishes movement to Britain and other places, giving in each case numbers inward and outward. The series on cross-border passenger movement distinguishes those travelling by rail and those travelling by scheduled omnibus road services. Travellers by private car are not included. Inward and outward movements are separately documented.[10] In addition, the Statistical Abstract presents *monthly traffic returns for all air routes into and out of Ireland.* This series documents passengers, freight and mail carried to and from Ireland by this mode. The staff strengths of Aer Lingus and Aerlínte Éireann (combined) are recorded in the Statistical Abstract. Five categories of employee are distinguished and the numbers of each class employed in Ireland/Europe, the USA and Britain are separately identified.

Expenditure on the maintenance and improvement of public roads is recorded in the Statistical Abstract. The amount of expenditure under each heading is shown by class of road and by the type of local authority undertaking it for each county. Total expenditure on public roads is documented and the separate contributions of local funds and state grants are identified.

A considerable volume of research has been carried out into transportation in Ireland. Transport accounted for 22% of gross energy consumption in Ireland in 1977, according to Feeney (1980A). This figure is high by EEC standards: it reflects the widespread use of cars in rural areas. Not surprisingly, transport is a significant user of imported energy, representing 30% of total oil consumption. Over 75% of such oil consumption is accounted for by road transport, of which almost two-thirds is accounted for by private motorists. Rail transport by contrast accounts for a mere 3% of total oil consumption. The elasticity of demand for fuel has been investigated by O'Riordan (1972) and by Feeney (1976). Both authors suggest that demand is price inelastic in both the short and the long-run. Projections of car ownership at county level for 1985 and 1995 have been carried out by McCarthy (1974) and at national level by O'Farrell et al. (1971).

The Irish transport system is both inadequate and inefficient, according to available evidence. Significant upgrading of the road network, in particular, may be required if rapid industrial growth is to be sustained. To this end the employment content of investment in roadworks has been analysed by Feeney (1980B). Using project cost data supplied by local authorities, his results suggest that at 1977 prices each million pounds expended on the road network created 134 man years of work. The employment thus generated is classified into four categories: direct on-site, direct off-site, indirect and induced. Such a high employment multiplier reflects in part the low import content of construction.

Communications

The communications sector is poorly documented compared to the wide, if somewhat disparate, range of data available on transport. The only regularly published data are to be found in the Statistical Abstract and refer to the number of television licences current in each county on 31 December of each year and an analysis of broadcasting on both radio and television by Radio Telefís Éireann. The number of hours broadcast is analysed by the type of programme and whether the output is home-produced or imported.

The post and telecommunications system is documented in the *Report of the Posts and Telegraphs Review Group 1978-1979.*[11] Chapter 4 of this report reviews the present state of the telecommunications system. Some useful statistics are provided on the size of the network and on the available standard of service in the final quarter of 1978. The report documents in the case of the telephone and telex services the degree of automation, the dialling success rate for both local and STD calls, the operator response time when called, the extent of faults in the system and the time taken to repair them, and the waiting time for installation of a telephone or telex machine. Separate data are provided for each of the Department of Post and Telegraphs Engineering Districts. Chapter 10 of the report analyses the present state of the postal system, documenting the average delivery times of mail. Appendix 2 to the report contains the results of a survey commissioned by the Review Group on the quality of the telephone and telex service in Ireland, while Appendix 3 consists of a study report on the Republic's telecommunications systems by the Bell System Telephone Company of the United States. Both Appendixes contain useful supplementary data on the state of the telephone system in 1978.

The only other source of data on communications is the *Household Budget Survey* of 1973 and the continuing urban Household Budget Surveys of later years which document the percentage of households having a telephone. In the case of the 1973 survey, the results are regionally disaggregated.

Notes to Chapter 9

[1] Note, however, that the catering activities of hotels, guesthouses and similar establishments were not included. The scope of vehicle hire services was defined to exclude taxis.

[2] The methodology whereby estimates were compiled for non-respondents is set out in the *Census of Distribution, 1971,* Summary Results for Retail and Wholesale Trade pp. v-xix. PRL 4709 Stationery Office, Dublin.

[3] Comparison between 1966 and 1971 censuses of distribution is necessarily limited to aggregates for which estimates for non-respondents were made on both occasions. No

such estimates were made for non-respondent wholesale establishments in 1966, while for retail trade they were made only for broad categories of turnover and employment. Comparison is further hindered by changes in the composition of turnover of individual establishments which may lead the same establishment to be assigned to different trading categories in successive censuses. This problem appears to be particularly acute in the case of drapery shops, department stores and general stores.

[4]   In particular, auctioneers and estate agents, bookmakers, and cinemas and theatres were not included in the 1971 census.

[5]   *'Sample survey of road freight transport 1964'* PRL 9572, Stationery Office, Dublin.

[6]   The Scope of this series was broadened in 1969. See: *'Road freight transport by licensed hauliers'* (Revised Series), ISB, December 1971, pp. 230-31, and *Irish Trade Journal and Statistical Bulletin,* December 1953.

[7]   *'National roads and traffic flows, 1977'* Paper RT 200, An Foras Forbartha, Dublin.

[8]   J. Devlin *'Vehicle kilometres of travel, 1979'* Paper RT 237, An Foras Forbartha, Dublin.

[9]   Traffic involving Lancashire flats, and enclosed or bulk liquid containers above a specified minimum size. Units mounted on goods vehicles or trailers on roll on-roll off vessels are not included.

[10]   The data on net passenger movement thus derived formed the basis of the CSO estimates of net migration to and from Ireland prior to 1979. See Hughes (1977, 1980) and Chapter 1 above.

[11]   *'Report of Posts and Telegraphs Review Group 1978-1979'* PRL 7883, Stationery Office, Dublin.

10 Regional Statistics

'It is considered essential to have established as soon as possible standard statistical regions for the compilation of regional statistics. This should be done immediately as a first step towards the provision of a range of regional statistics.' *Report of the Committee of Statistical Requirements and Priorities,* 1974.

There can be little disagreement with this recommendation, although it has gone largely unnoticed. Standard statistical regions have not been identified nor has anything remotely approaching a comprehensive system of regional statistics been designed. This is regrettable as an analysis of available regional statistics (Kirwan and McGilvray, 1980) suggests that Ireland is characterised by a high degree of regional heterogeneity. It is a measure of the lack of progress in the provision of regional statistics that Curtin's (1972) review of available data still remains reasonably accurate.

Insofar as the CSO intends to produce regional statistics, the areal units used will probably be the nine IDA planning regions. These regions are simply aggregations of counties. Table 10.1 and the accompanying map provide the details. Given that many series are available on a county basis, regional statistics thus defined should be relatively cheap to provide and, in view of the low number of regions, should minimise problems of confidentiality.

Table 10.1 IDA planning regions

Region	Area
East	Dublin, Kildare, Meath and Wicklow
South-East	Carlow, Kilkenny, Tipperary SR, Wexford, Waterford
South-West	Cork and Kerry
Mid-West	Clare, Limerick and Tipperary NR
West	Galway and Mayo
North-West	Leitrim and Sligo

Figure 10.1 IDA planning regions

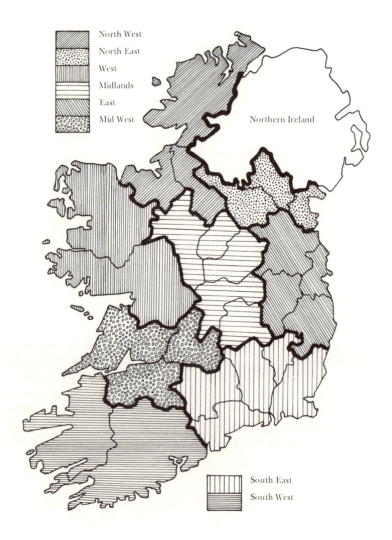

Donegal   Donegal
Midlands   Laois, Longford, Offaly, Roscommon and Westmeath
North-East   Cavan, Louth and Monaghan

This chapter reviews the range of economic statistics which have either been provided on the planning region basis or which can be aggregated to this level from existing published data. The extent to which regional statistics are available on each subject heading of previous chapters will be reviewed in turn.

Population and vital statistics

The main data sources under this heading are the decennial *Census of Population,* the biennial *Labour Force Survey* and the *Report on Vital Statistics.* However, the Labour Force Survey is the only source to aggregate its results according to planning region, though some of the summary results of the 1971, 1979 and 1981 censuses have also been provided on a regional basis. Most of the remaining results are at a lower, generally county, level of aggregation. It is possible therefore to prepare a series on regional population totals going back over a considerable period by simply aggregating the county results of previous censuses of population. Regional population estimates for non-census years can be gleaned from the results of the Labour Force Survey, but (as noted in Chapter 1) these figures will not be strictly consistent with those of the census. The survey-based estimates suffer from a number of drawbacks and are liable to revision in the light of subsequent census results. Furthermore, the CSO advises caution in comparing regional data from successive surveys because of changes in survey methodology and because of the large standard errors inherent in any comparison of estimates from independent samples.[1] The figures from either source may be disaggregated by age, sex and marital status.

The *Household Budget Survey* of 1973 provides regional estimates of the age and sex composition of households stratified by household income, household tenure, social group, livelihood status of principal breadwinner and size of household.

A wealth of spatially disaggregated raw data is available on a regular basis on the subjects of mortality and fertility. There is however an almost total absence of regular data on migration which as noted above is the most volatile component of population change. The main published source of interest here is the annual Report on Vital Statistics. This publication appears however with a lag of three to four years. Considerably more aggregated statistics are published in the *Quarterly Report on Births, Deaths, Marriages and Certain Infectious Diseases,* where the publication lag is about three months. Neither of these publications presents any of its results on a planning region basis. However, most of the important fertility and mortality indicators can be derived on this basis, if somewhat tediously, from the published results.

Crude birth rates for each of the planning regions can be readily computed. Given an estimate of the population of women of child bearing age, derived from either the Census or the Labour Force Survey, general fertility rates may be compared across regions. Age-specific fertility rates of broad age bands may be calculated at a regional level given the Labour Force Survey population estimates or at a more disaggregated level using Census of Population results. In either case these age-specific rates may be defined in terms of either all, or simply married, women. The relative importance of births to unmarried mothers may also be compared by planning region. Infant mortality is well-documented at the county level, distinguishing deaths of infants aged under four weeks and those aged between four weeks and one year. Rates of neonatal and infant mortality may therefore be computed for individual regions; further disaggregation by cause of death is possible.

The number of deaths assigned to each county, classified by age, sex and cause of death, is readily documented in the Report on Vital Statistics and the Quarterly Report on Births, Deaths, Marriages and Certain Infectious Diseases. Given appropriately disaggregated regional population estimates — whether from the Census of Population or the Labour Force Survey — crude, standardised, and age-specific death rates may be calculated with further disaggregation by cause of death possible.

Net and gross inter-regional migrant flows may be derived from the Census and Labour Force Survey results.

Manpower

The principal sources of data on the regional structure of employment are the decennial Census of Population, the annual *Census of Industrial*

Production, and *Census of Distribution* and, for recent years, the Labour Force Survey.

Benchmark data on the sectoral and occupational distribution of employment classified by sex within each county are provided by the Census of Population results; and corresponding regional totals may be derived from these. The regional distribution of white collar, technical and professional employment in 1961 and 1971 has been analysed by NESC (1977) using such data while employment relationships at county level have been examined by Baker and Ross (1975). Similar regional data (at a somewhat more aggregate sectoral and occupational level) are available from the results of the Labour Force Survey, though (as noted above) results from successive surveys should be compared with caution. These are the only available sources of data on total employment by region. The remaining sources provide disaggregated information only on specific sectors.

For years prior to 1973 the Annual Census of Industrial Production yields county data on employment in census industries. These data are to be found in the Statistical Abstract.

The AnCo *Manpower Surveys* cover only manufacturing and building and construction. Detailed results are presented by planning region for eight industrial sectors, providing for each sector in a region a breakdown of employment by broad occupational group. Within each sector the occupational data for each region are further disaggregated by major industrial headings. Information is also provided on the regional distribution of apprentices by occupation within certain industries.

The IDA *Employment Survey* provides data on the regional distribution of manufacturing employment by sex. No disaggregation of the results by age, occupation or industry is available. The *Annual Reports* of the IDA document employment creation in each region and the amounts of grant aid and financial assistance expended to this end.

The annual *Agriculture Enumeration* provides county estimates of the number of males engaged either permanently or temporarily in farm work. Those engaged are classified as 'family members' or 'others', distinguishing those aged fourteen to eighteen, and those aged eighteen and over. No data are provided on female employment in the agricultural sector, though a regional estimate of this is available from the Labour Force Survey.

Nearly all published statistics on unemployment in Ireland are a by-product of the administration of schemes of unemployment compensation. In order to claim such compensation it is necessary to register at a local office of the Department of Social Welfare. Accordingly, most unemployment series are available at the level of local employment offices. It is therefore possible to distinguish for each office the number of persons who register as unemployed in each month (called the live register) disaggregated by industrial group, age, occupation and sex. These series can readily be aggregated to give corresponding planning region totals, and since 1978 the CSO has published the results on such a basis. Manual aggregation is required for earlier years. A regionally disaggregated analysis of the live register is also to be found in the *Manpower Information Quarterly*.

Alternative measures of unemployment can be derived from the results of the Census of Population and Labour Force Surveys. The Labour Force Survey is of particular interest as it distinguishes three categories of unemployed person; not all of these would be recorded on the live register. Since 1977 the categories distinguished are:

1.   unemployed, having lost or given up previous job
2.   looking for first regular job
3.   unable to work, due to permanent sickness or disability

Estimates, disaggregated by sex, of the number in each category are provided for individual regions. It is thus possible, using data on principal economic status from the same source, to calculate unemployment rates by sex for each region for Labour Force Survey years. The resulting estimates will differ from published unemployment rates which measure the percentage of the insured population who are unemployed. At present the latter are published on a quarterly basis for each of the planning regions in the annual *Trend of Employment and Unemployment*. The coverage of these figures is considerably narrower than that of the Labour Force Survey.

Agriculture

The results of the annual *Census of Agriculture* and the various livestock enumerations are generally published at the county level. It is therefore possible to aggregate without difficulty to planning regions. The following statistics can be readily compiled at this level:

(1)   The area under various types of crops
(2)   The number of cattle, sheep, pigs, poultry and horses
(3)   The number of males engaged in farm work classified by age,

permanent or temporary employment, family member or other.

A potentially important source of regional data on the agricultural sector is the annual *Farm Management Survey* from An Foras Talúntais. The results of this survey have been aggregated to planning region level since 1978. Data for farms participating in the survey are classified in two ways. The first is by soil type: three types are distinguished; and the second is by farming system: four systems are distinguished. The results of the survey provide detailed information on the inputs and outputs of each farm; these flows are also standardised by the cultivable area of the farm. The gross output and labour income of each farm are also standardised by the number of units of labour input employed. A variety of summary figures is presented including the size of the farm in acres and the size in acres adjusted for any uncultivable land, actual investment in machinery and livestock, stocking densities and the breakdown of labour units employed into family and hired labour.

Industrial production

Almost no regionally disaggregated data on industrial production are published. Before 1973, the annual *Statistical Abstract* contained details of output and employment for the industrial sector in each county. These can without difficulty be re-aggregated to regional level. The figures presented are for the whole of the industrial sector, and no disaggregation by industrial group is possible. Only twice, in special analyses of the 1968 and 1975 Censuses of Industrial Production, have these results been regionally disaggregated by broad industrial sub-groups.

Foreign trade and national income

No attempt has been made to compile this data on a regional basis. In large economies with more pronounced regional problems, data on regional trade flows and regional accounts contribute to the analysis of interregional differentials. It is debatable whether this data deserves a high priority in Ireland at present.

Income, wealth and expenditure

The work of Attwood and Geary (1963), Ross (1969, 1972, 1980), Ross and Jones (1977) provides estimates of gross personal incomes by county 1960, 1965, 1969, 1973 and 1977. Ross and Jones (1977) also include corresponding estimates of regional incomes, while the results of earlier years may be readily aggregated to this level. The figures can be converted to a per capita basis by deflating by the size of regional population. Certain caveats are in order, however, if these data are to be used for purposes of interregional comparison. The figures are in current prices, so that deflation by a price index is necessary if real incomes are to be compared. However, only a national price index is available for this purpose, and so no account can be taken of regional variations in price and consumption vectors. Furthermore, the income estimates are gross figures, while the most relevant concept for interregional comparison is that of disposable income. Direct taxation absorbs a considerably smaller proportion of agricultural than of non-agricultural incomes. As a result, interregional comparisons of gross incomes may exaggerate the true degree of regional inequality. The practice of valuing farmers' consumption of their own produce at farm-gate rather than at retail prices further heightens the apparent inequality.

The Household Budget Survey of 1973, the first to include both urban and rural households, provides estimates of regional weekly household income. The incomes of the responding households in each region are disaggregated by source, distinguishing seven categories of direct income and six categories of state transfers. Income tax and social insurance deductions are also identified, enabling an estimate to be made of household disposable income by region. The results are further disaggregated by household size, household tenure, four bands of average household income and the economic status of the household head. These disaggregated results have been used by Nolan (1978A) in an attempt to estimate the distribution of income within each region.

There are no officially published statistics on the regional distribution of wealth in Ireland. Lyons (1972) has attempted to estimate such a distribution at county level using data on the characteristics of wills submitted for probate. The methodology employed attracted considerable criticism, however (Chesher and McMahon (1976), Harrison and Nolan (1975), Chesher (1979)). Proxy measures of personal wealth at the regional level can be derived from the 1973 Household Budget Survey which documents the proportion of households possessing specified domestic appliances, specified forms of central heating and one or more cars.

Regional expenditure patterns are also detailed in the Household

Budget Survey. Average weekly household expenditure on a wide variety of goods and services for four different household income levels is documented for each region. Slightly more aggregated expenditure data is also presented with a further classification by household tenure, livelihood status of the head of the household, size of household and social group.

Very little systematic data exist on the regional distribution of central government expenditure (Kirwan and McGilvray, 1980). By contrast, the current expenditure and income of local authorities is well-documented in both the annual *Returns of Local Taxation* and its companion volume, *Local Authority Income and Expenditure*. The economic aspects of such expenditure at the county level have been examined by Copeland and Walsh (1975).

Prices and wages

There are practically no published statistics at either regional or county level of prices and wages in any sector of the economy. The exceptions relate to average prices of grocery and domestic non-durable goods in particular towns published by the National Price Commission, and to house prices, where average new and secondhand prices for seven major counties are published by the Department of the Environment in the *Quarterly Bulletin of Housing Statistics*.

Distribution, transport and communication

The 1971 Census of Distribution provided a certain amount of information on the characteristics of wholesale and retail outputs at the regional level. By contrast, very little regionally disaggregated data are available in the fields of transport and communication. The *Statistical Abstract* contains details of the number of mechanically-propelled vehicles under current licence, distinguishing various classes of private, public and commercial transport by county of registration. The abstract also documents the number of driving licences current in each county. The report of the Posts and Telegraphs Review Group (1979) provides a certain amount of regional data on communications. The data are not based on planning regions, but on the engineering districts of the Department of Posts and Telegraphs. Unfortunately, these districts do not coincide with the planning regions. The report documents the average waiting time for installation of telephone, the percentage of telephone lines which are automatic and the STD dialling success rate to and from Dublin for each district during 1978.

Concluding comments

Regional policy has been the subject of considerable debate in Ireland since the early 1960s (NESC 1975C, 1976B, 1977). However, the development of a regionally disaggregated economic data base to give substance to the debate has not been forthcoming. As noted at the start, the report of the Committee on Statistical Requirements and Priorities (1974) recommended the development of such an integrated system of regional statistics. To date, progress has been exceedingly slow.

Notes to Chapter 10

[1]   See *'Labour Force Survey, 1979 Results'*, for a detailed comparison of the results of the first three surveys.

Bibliography

Atkinson, A.B. (1974) 'Poverty and income inequality in Britain' in *Poverty, inequality and class structure* ed D. Wedderburn London: Cambridge University Press

Attwood, A. and Geary, R.C. (1963) 'Irish county incomes in 1960' *ESRI General Research Series, Paper No 16*

Baker, T.J. (1966) 'Regional employment patterns in the Republic of Ireland' *ESRI General Research Series Paper No 32*

Baker, T.J. and Neary, J.P. (1971) 'A study of consumer prices, Part 1' *ESRI Quarterly Economic Commentary* March 1971

Baker, T.J. and Ross, M. (1975) 'Employment relationships in Irish counties' *ESRI General Research Series Paper No. 81*

Black, W., Simpson, J.V. and Slattery, D.G. (1969) 'Costs and prices in transportable goods industries' *ESRI General Research Series, Paper No 51*

Bradley, J. (1977) 'Seasonality and unemployment in Ireland: comments' Central Bank of Ireland, mimeo

Bradley, J. (1978) 'Personal income taxation and inflation in Ireland 1971-77' *Journal of the Statistical and Social Inquiry Society of Ireland* Vol XXIII, Part V, pp. 277-96

Chambers, A.M. (1974) 'Land price trends in Ireland and selected EEC countries' *Central Bank of Ireland Quarterly Bulletin No. 4.*

Chesher, A.D. and McMahon, P.C. (1976) 'The distribution of personal wealth in Ireland — the evidence re-examined' *Economic and Social Review* Vol 8, No 1, pp. 61-5

Chesher, A.D. (1979) 'An analysis of the distribution of wealth in Ireland' *Economic and Social Review* Vol 11, No 1, pp. 1-18

Cogan, D.J. (1978) 'The Irish services sector — a study of productive efficiency' Stationery Office

Connell, K.H. (1950) *The population of Ireland 1750-1845* Oxford University Press, Oxford

Conniffe, D. and Killen, L. (1977) 'Irish consumer expectations and the adaptive expectations model' *Economic and Social Review* Vol 9, No 1, pp. 71-7

Copeland, J.R. and Walsh, B.M. (1975) 'Economic aspects of local authority expenditure and finance' *ESRI General Research Series, Paper No 84*

Cowling, Keith (1966) 'Determinants of wage inflation in Ireland' *ESRI General Research Series, Paper No 31*

Curtin, J.V. (1972) *A guide to regional statistics* An Foras Forbartha

de Buitleir, D. (1974) *Problems of Irish local finance* Institute of Public Administration

Devlin, J. (1980) 'Vehicle kilometers of travel, 1979' *Paper RT 237* An Foras Forbartha

Dowling, B.R. (1975) 'Seasonality and unemployment in Ireland'. *ESRI Quarterly Economic Commentary* October 1975, pp. 37-44

Dowling, B.R. (1977) 'The income sensitivity of the personal income tax base in Ireland 1947-1972' *ESRI General Research Series Paper No 86*

Dowling, B.R. (1978) 'Integrated approaches to personal income taxes and transfers' *NESC Report No 37,* NESC

Feeney, B.P. (1976) 'The demand for petrol' *Paper RT 162,* An Foras Forbartha

Feeney, B.P. (1980A) 'Energy and road transportation — the need for research' *Paper RT 230,* An Foras Forbartha, Dublin

Feeney, B.P. (1980B) 'Employment content of roadworks' *Paper RT 193,* An Foras Forbartha

Fraser, N.A. and Moar, L. (1981) 'Compilation, analysis and updating, of occupation by industry matrices for Scotland 1961-1977' Fraser of Allander Institute, University of Strathclyde, mimeo

Geary, P.T. and Jones, R.M. (1975) 'The appropriate measure of unemployment in an Irish Phillips Curve' *Economic and Social Review* Vol 7, No 1, pp. 55-64

Geary, P.T., Walsh, B.M. and Copeland, J. (1975) 'The cost of capital to Irish industry' *Economic and Social Review* Vol 6, No 3, pp. 299-312

Geary, P.T. (1976A) 'Lags in the transmission of inflation: some preliminary estimates' *Economic and Social Review* Vol 7, No 4, pp. 383-9

Geary, P.T. (1976B) 'World prices and the inflationary process in a small open economy — the case of Ireland' *Economic and Social Review* Vol 7, No 4, pp. 391-400

Geary, P.T. (1977) 'Wages, prices, incomes and wealth' in *Economic activity in Ireland: a study of two open economies* eds N.J. Gibson and J.E. Spenser, Gill and Macmillan

Geary, P.T. and McCarthy, C. (1977) 'Wage and price determination in a labour-exporting economy: the case of Ireland' *European Economic Review* Vol 8, pp. 219-33

Geary, P.T. and McDonnell, E. (1979) 'The cost of capital to Irish industry: revised estimates' *Economic and Social Review* Vol 10, No 4, pp. 287-300

Geary, R.C. and Hughes, J.G. (1970) 'Internal migration in Ireland' *ESRI Paper No 54*

Geary, R.C. (1973) 'Quarterly non-agricultural stock statistics: a pilot inquiry' *ESRI Quarterly Economic Commentary* January 1973

Glass, D.V. and Taylor, P.A.M. (1976) *Population and emigration* Irish University Press

Harrison, M.J. and Nolan, S. (1975) 'The distribution of personal wealth in Ireland: a comment' *Economic and Social Review* Vol 7, No 1, pp. 65-78

Hughes, J.G. (1972) 'The functional distribution of income in Ireland 1938-70' *ESRI General Research Series Paper No 65*

Hughes, J.G. and Walsh, B.M. (1975) 'Migration flows between Ireland, the United Kingdom and the rest of the world 1966-1971 *European Demographic Information Bulletin* Vol 7, pp. 125-49

Hughes, J.G. (1977) 'Estimates of annual net migration and their relationship with series on annual net passenger movement: Ireland 1926-1976' *ESRI Memorandum Series No 122*

Hughes, J.G. (1980) 'What went wrong with Ireland's recent postcensal population estimates' *Economic and Social Review* Vol 11, No 2, pp. 137-46

Hughes, J.G. and Walsh, B.M. (1980) 'Internal migration flows in Ireland and their determinants' *ESRI General Research Series Paper No 98*

Keating, W. (1977) 'An analysis of recent demographic trends with population projections for the years 1981 and 1986' *Journal of the Statistical and Social Inquiry Society of Ireland* Vol XXIII, Part IV, pp. 113-151

Keenan, J.G. (1978) 'Unemployment, emigration and the labour force' in *Irish economic policy: a review of major issues* eds B.R. Dowling and J. Durkan ESRI

Keenan, J.G. (1981) 'Irish migration, all or nothing resolved?' *Economic and Social Review* Vol 12, No 3, pp. 169-86

Kelleher, R. (1977) 'The influence of liquid assets and the sectoral distribution of income on aggregate consumers' behaviour in Ireland' *Economic and Social Review* Vol 8, No 3, pp. 187-200

Kennedy, K.A. and Dowling, B.R. (1970) 'The determinants of personal savings in Ireland: an econometric enquiry' *Economic and Social Review* Vol 2, No 1, pp. 19-51

Kennedy, K.A., Walsh, B.M. and Ebrill, L. (1973) 'The demand for beer and spirits in Ireland' *Proceedings of the Royal Irish Academy* Vol 73, Section C, Number 13

Kennedy, K.A. and Bruton, R. (1975) 'The Consumer Price Index and different household expenditure patterns' *ESRI Quarterly Economic Commentary,* pp. 27-36

Kennedy, K. and Dowling, B.R. (1976) *Economic growth in Ireland* Gill and Macmillan

Kirwan, F.X. (1979) 'Non wage costs, employment and hours of work in Irish manufacturing industry' *Economic and Social Review* Vol 10, pp. 231-54

Kirwan, F.X. and McGilvray, J.W. (1980) 'The identification and construction of key social indicators for use in regional planning and development in Ireland' *Report to the Department of Finance*

Kirwan, F.X. (1982) 'Recent Anglo-Irish migration — the evidence of the British labour force surveys' *Economic and Social Review* Vol 13, No 2.

Lee, J. (1973) ed *The population of Ireland before the 19th century* Gregg International Publishers, London

Lennan, L.K. (1972) 'The built-in flexibility of Irish taxes' *Economic and Social Review* Vol 3, No 4, pp. 581-604

Leser, C.E.V. (1962) 'Demand relationships for Ireland' *ESRI General Research Series Paper No 4*

Leser, C.E.V. (1964) 'A further analysis of Irish household budget data 1951-1952' *ESRI General Research Series Paper No 23*

Lyons, P.M. (1972) 'The distribution of personal wealth in Ireland' in *Ireland: some problems of a developing economy* eds A.A. Tait and J.A. Bristow, Gill and Macmillan, Dublin; New York: Barnes and Noble

Lyons, P.M. (1973) 'The distribution of personal wealth by county in Ireland, 1966' *Journal of the Statistical and Social Inquiry Society of Ireland* Vol XXII, Part V, 1972-1973, pp. 78-109

Lyons, P.M. (1974) 'The size distribution of personal wealth in Ireland' *Review of Income and Wealth* Vol 20, No 2

Lyons, P.M. (1975) 'Estate duty wealth estimates and the mortality multiplier' *Economic and Social Review* Vol 6, No 3, pp. 337-352

McCarthy, C. (1974) 'Car number projections by county for 1985/1995' *Paper RT 115,* An Foras Forbartha, Dublin

McCarthy, C. and Ryan, J. (1976): 'An econometric model of television ownership' *Economic and Social Review* Vol 7, No 3, pp. 265-78

McCarthy, C. (1977A) 'Estimates of a system of demand equations using alternative commodity classifications of Irish data, 1953-1974' *Economic and Social Review* Vol 8, No 3, pp. 201-12

McCarthy, C. (1977B) 'An econometric model of non-agricultural stock changes' *ESRI Quarterly Economic Commentary* pp. 20-28

Murphy, M.J. (1973) 'The CSO mixed model for seasonal adjustment and model test programs' *British Central Statistical Office Research Exercise Note 5/73*

Neary, J.P. (1975) 'The CII-ESRI. Quarterly and monthly surveys of business attitudes: methods and uses' *ESRI Quarterly Economic Commentary* March 1975

NESC (1975A): 'Report on inflation' *National Economic and Social Council Report, No 9* Stationery Office

NESC (1975B): 'Causes and effects of inflation in Ireland' *National Economic and Social Council Report, No 10* Stationery Office

NESC (1975C): 'Regional policy in Ireland: a review' *National Economic and Social Council Report, No 4* Stationery Office

NESC (1975D): 'Income distribution: a preliminary report' *National Economic and Social Council Report, No 11* Stationery Office

NESC (1975E): 'Jobs and living standards: projections and implications' *National Economic and Social Council Report No 7* Stationery Office

NESC (1976A): 'Population projections 1971-86: the implications for education' *National Economic and Social Council Report No 18,* Stationery Office

NESC (1976B): 'Institutional arrangements for regional economic development' *National Economic and Social Council Report, No 22* Stationery Office

NESC (1976C): 'Population projections 1971-1986: the implications for social planning – dwelling needs' *National Economic and Social Council Report No 14* Stationery Office

NESC (1977) 'Service-type employment and regional development' *National Economic and Social Council Report, No 28*

Nevin, E. (1962A): 'The ownership of personal property in Ireland' *ESRI General Research Series Paper No 1*

Nevin, E. (1962B): 'Wages in Ireland 1946-62' *ESRI General Research Series Paper No 12*

Nolan, B. (1978A) 'The personal distribution of income in the Republic of Ireland' *Journal of the Statistical and Social Inquiry Society of Ireland* Vol XXIII, Part V, 1977-8, pp. 91-162

Nolan, B. (1978A) 'The personal distribution of income in the Republic Program' *Central Bank of Ireland Annual Report* pp. 86-98

Norton, D. (1976) 'Income distribution in the Republic of Ireland: a note' *Social Studies* July 1976, pp. 66-74

Norton, D. and O'Donnell, R. (1978) 'The Irish PAYE income tax system' *Economic and Social Review* Vol 9, No 2, pp. 107-24

O'Connor, R. and Henry, E.W. (1975) 'Input-output analysis and its applications' *Griffins Statistical Monograph and Courses* No 36, London

O'Farrell, P.N., Markham, J. and McLoughlin, T. (1971) 'A note on national and county car ownership projections in Ireland' *Economic and Social Review* Vol 3, No 1, pp. 95-105

O'Farrell, P.N. (1970A) 'Regional development in Ireland: the economic case for regional policy *Administration* Vol 18, No 4

O'Farrell, P.N. (1970B) 'Regional development in Ireland: problems of goal formulation and objective specification *Economic and Social Review* Vol 2, No 1, pp. 71-92

O'Farrell, P.N. (1972A) 'The regional problem in Ireland: some reflections upon development strategy' *Economic and Social Review* Vol 2, No 4, pp. 453-80

O'Farrell, P.N. (1972B) 'A shift and share analysis of regional employment change in Ireland 1951-1966' *Economic and Social Review* Vol 4, No 1, pp. 59-86

O'Farrell, P.N. (1974) 'Regional planning in Ireland — the case for concentration: a reappraisal' *Economic and Social Review* Vol 5, No 4, pp. 499-514

O'Farrell, P.N. (1975) *Regional industrial development trends in Ireland 1960-1973* Institute of Public Administration for Industrial Development Authority, Dublin

O'Grada, C. (1975) 'A note on nineteenth century Irish emigration statistics' *Population Studies* Vol 29, pp. 143-9

O'Muircheartaigh, F.S. (1977) 'The changing burdens of personal income tax in Ireland and the social valuation of income 1946-1976' *Journal of the Statistical and Social Inquiry Society of Ireland* Vol XXIII, Part IV, 1976-7

O'Reilly, L. (1981) 'Estimating quarterly national accounts' *Proceedings of the Statistical and Social Inquiry Society of Ireland* November 1981

O'Reilly, L. and Gray, A. (1980) 'Seasonality and other components of the Irish Unemployment Series' *ESRI Quarterly Economic Commentary,* April 1980, pp. 37-54

O'Riordan, W.K. (1972) 'The elasticity of demand for petrol in Ireland' *Economic and Social Review* Vol 3, No 3, pp. 475-85

O'Riordan, W.K. (1975) 'An application of the Rotterdam demand system to Irish data' *Economic and Social Review* Vol 6, No 4, pp. 511-32

O'Riordan, W.K. (1976) 'The demand for food in Ireland 1947-1973' *Economic and Social Review* Vol 7, No 4, pp. 401-16

Paglin, P. (1975) 'The measurement of inequality: a basic revision' *American Economic Review* Vol LXV, pp. 598-609

Pratschke, J.L. (1970) 'Income – expenditure relations in Ireland 1965-1966' *ESRI General Research Series Paper No 50*

Reason, L. (1961) 'Estimates of the distribution of non-agricultural incomes and incidence of certain taxes' *Journal of the Statistical and Social Inquiry Society of Ireland* Vol XX, Part IV, 1960-1

Report of the Committee on Statistical Requirements and Priorities (1974) Stationery Office, Dublin

Report of the Inter-Departmental Committee on Unemployment Statistics (1979) Stationery Office

Report of Posts and Telegraphs Review Group (1979) Stationery Office, Dublin

Ross, M. (1969) 'Personal incomes by county 1965' *ESRI General Research Series Paper No 49*

Ross, M. (1971) 'Methodology of personal income estimation by county' *ESRI General Research Series Paper No 63*

Ross, M. (1972) 'Further data on county incomes in the sixties' *ESRI General Research Series Paper No 64*

Ross, M. and Jones, R. (1977) 'Personal incomes by county in 1973' *NESC Report No 30*

Ross, M. and Walsh, B.M. (1979) 'Regional policy and the full employment target' *ESRI Policy Research Series No 1*

Ross, M. (1980) 'Personal incomes by region in 1977' *NESC Report No 51,* Dublin

Ruane, F. (1975) 'A statistical analysis of revisions to the Irish national accounts' *Economic and Social Review* Vol 6, No 3, pp. 387-404

Samuelson, P.A. (1976) *Economics: an introductory analysis,* McGraw-Hill, London, 10th edition

Sandell, S.H. (1974) 'Non-agricultural unemployment and its measurement in Ireland' *Central Bank of Ireland Annual Report 1974,* pp. 110-126

Shiskin, J., Young, A.H. and Musgrave, J.C. (1965) 'The X-11 variant of the census method II seasonal adjustment program' *Technical Paper No 15* Bureau of the Census, US Department of Commerce

Shorter, F.C. and Pasta, D. (1974) *Computational Methods for Population Projections* The Population Council, New York

Taylor, J. (1974) *Unemployment and wage inflation* Longman, London

Thompson, W.J. (1911) 'The development of the Irish census and its national importance' *Journal of the Social and Statistical Inquiry Society of Ireland* Part 91

United Nations (1968) *A system of national accounts* Series F, No 2, Rev 3, New York

Vaughan, W.E. and Fitzpatrick, A.J. (1978) (eds) *Irish historical statistics: population 1821-1971* Royal Irish Academy (Dublin)

Vaughan, R.N. (1980) 'Measures of the capital stock in the Irish manufacturing sector, 1945-1973' *ESRI General Research Series Paper No 103,* Dublin

Walsh, B.M. (1968) 'Some Irish population problems reconsidered' *ESRI General Research Series Paper No 42*

Walsh, B.M. (1972) 'Ireland's demographic transformation 1958-70' *Economic and Social Review* Vol 3, No 2, pp. 251-75

Walsh, B.M. (1974A) 'Expectations, information and human migration: specifying an econometric model of Irish emigration to Britain *Journal of Regional Science* Vol 14, No 1, pp. 107-20

Walsh, B.M. and Whelan, B.J. (1974) 'The determinants of female labour force participation: an econometric analysis of survey data' *Journal of the Statistical and Social Inquiry Society of Ireland* Vol XXIII, Part 1, 1973-4, pp. 1-33

Walsh, B.M. (1974B) 'The structure of unemployment in Ireland, 1954-1972' *ESRI General Research Series No 77*

Walsh, B.M. (1975) 'Population and employment projections 1971-86' *National Economic and Social Council Report No 5,* Stationery Office, Dublin

Walsh, B.M. (1977A) 'Unemployment, vacancies, and "full employment" in the Irish manufacturing sector' *ESRI Quarterly Economic Commentary* June 1977, pp. 25-35

Walsh, B.M. (1977B) 'Population and employment projections 1986: a reassessment' *NESC Report No 35*

Walsh, B.M. (1978) 'Unemployment compensation and the rate of unemployment: the Irish experience' in *Unemployment insurance* eds Grubel, H.G. and Walker, M.A., Fraser Institute, Vancouver

Walsh, B.M. (1980A) 'National and regional demographic trends' *Administration* Vol 26, No 2

Walsh, B.M. (1980B) 'Health education and the demand for tobacco in Ireland: a note' *Economic and Social Review* Vol 11, No 2, pp. 147-51

Whelan, B.J. and Walsh, B.M. (1977) 'Redundancy and re-employment in Ireland' *ESRI General Research Series No 89*

Whelan, B.J. and Keogh, G. (1980) 'The use of the Irish electoral register for population estimation' *Economic and Social Review* Vol 11, pp. 301-17

Whelan, B.J. (1977) 'RANSAM: a national random sample design for Ireland' ESRI mimeo

Yeomans, K.A. (1968) *Statistics for the social scientist* 2 Vols Vol 1: *Introducing statistics* Harmondsworth, Penguin

Index

**This book is to be returned on or before
the last date stamped below.**

Irish Economic Statistics

IRELAND

HOUSEHOLD BUDGET SURVEY

ANNUAL URBAN INQUIRY

RESULTS FOR 1979

COMPILED BY THE
CENTRAL STATISTICS OFFICE
DUBLIN

DUBLIN
PUBLISHED BY THE STATIONERY OFFICE

To be purchased from the
GOVERNMENT PUBLICATIONS SALE OFFICE, G.P.O. ARCADE, DUBLIN 1
or through any Bookseller

Pl. 111   Price £1.80   October 1981

CENSUS OF POPULATION

OF IRELAND

1981

VOLUME 1

POPULATION OF DISTRICT ELECTORAL DIVISIONS,
TOWNS AND LARGER UNITS OF AREA

COMPILED BY THE
CENTRAL STATISTICS OFFICE
The statistics herein relate to the population of Ireland exclusive of Northern Ireland

DUBLIN
PUBLISHED BY THE STATIONERY OFFICE

To be purchased from the
GOVERNMENT PUBLICATIONS SALE OFFICE, G.P.O. ARCADE, DUBLIN 1
or through any Bookseller

Price £3.70

Pl. 983   September, 1982.